AF834475

Effectiveness of the United States Advanced Battery Consortium as a Government-Industry Partnership

Committee to Review the U.S. Advanced Battery Consortium's
Electric Vehicle Battery Research and Development
Project Selection Process

Board on Energy and Environmental Systems

Commission on Engineering and Technical Systems

National Research Council

NATIONAL ACADEMY PRESS
Washington, D.C. 1998

NATIONAL ACADEMY PRESS • 2101 Constitution Avenue, N.W. • Washington, D.C. 20418

NOTICE: The project that is the subject of this report was approved by the Governing Board of the National Research Council, whose members are drawn from the councils of the National Academy of Sciences, the National Academy of Engineering, and the Institute of Medicine. The members of the committee responsible for the report were chosen for their special competences and with regard for appropriate balance.

The National Academy of Sciences is a private, nonprofit, self-perpetuating society of distinguished scholars engaged in scientific and engineering research, dedicated to the furtherance of science and technology and to their use for the general welfare. Upon the authority of the charter granted to it by the Congress in 1863, the Academy has a mandate that requires it to advise the federal government on scientific and technical matters. Dr. Bruce M. Alberts is president of the National Academy of Sciences.

The National Academy of Engineering was established in 1964, under the charter of the National Academy of Sciences, as a parallel organization of outstanding engineers. It is autonomous in its administration and in the selection of its members, sharing with the National Academy of Sciences the responsibility for advising the federal government. The National Academy of Engineering also sponsors engineering programs aimed at meeting national needs, encourages education and research, and recognizes the superior achievements of engineers. Dr. William A. Wulf is president of the National Academy of Engineering.

The Institute of Medicine was established in 1970 by the National Academy of Sciences to secure the services of eminent members of appropriate professions in the examination of policy matters pertaining to the health of the public. The Institute acts under the responsibility given to the National Academy of Sciences by its congressional charter to be an adviser to the federal government and, upon its own initiative, to identify issues of medical care, research, and education. Dr. Kenneth I. Shine is president of the Institute of Medicine.

The National Research Council was organized by the National Academy of Sciences in 1916 to associate the broad community of science and technology with the Academy's purposes of furthering knowledge and advising the federal government. Functioning in accordance with general policies determined by the Academy, the Council has become the principal operating agency of both the National Academy of Sciences and the National Academy of Engineering in providing services to the government, the public, and the scientific and engineering communities. The Council is administered jointly by both Academies and the Institute of Medicine. Dr. Bruce M. Alberts and Dr. William A. Wulf are chairman and vice chairman, respectively, of the National Research Council.

This report and the study on which it is based were supported by Contract No. DTNH22-94-G-07414 from the National Highway Traffic Safety Administration, U.S. Department of Transportation. Any opinions, findings, conclusions, or recommendations expressed in this publication are those of the author(s) and do not necessarily reflect the views of the organizations or agencies that provided support for the project.

International Standard Book Number: 0-309-06143-1
Library of Congress Catalog Card Number: 98-86227

Available in limited supply from:

Board on Energy and Environmental Systems
National Research Council
2101 Constitution Avenue, N.W.
HA-270
Washington, DC 20418
202-334-3344

Additional copies are available for sale from:

National Academy Press
2101 Constitution Avenue, N.W.
Box 285
Washington, DC 20055
800-624-6242 or 202-334-3313 (in the Washington metropolitan area)
http://www.nap.edu

Copyright 1998 by the National Academy of Sciences. All rights reserved.
Printed in the United States of America

COMMITTEE TO REVIEW THE U.S. ADVANCED BATTERY CONSORTIUM'S ELECTRIC VEHICLE BATTERY RESEARCH AND DEVELOPMENT PROJECT SELECTION PROCESS

LARRY R. FAULKNER (*chair*), University of Texas at Austin
KATHRYN R. BULLOCK, Medtronic, Inc., Brooklyn Center, Minnesota
PAUL A. KOHL, Georgia Institute of Technology, Atlanta
CARL A. KUKKONEN, California Institute of Technology, Pasadena
ALEXANDER MACLACHLAN, NAE,[1] E.I. du Pont de Nemours & Company (retired), Wilmington, Delaware
JAMES A. MCINTYRE, Dow Chemical Company, Midland, Michigan
BARRY MILLER, Case Western Reserve University, Cleveland, Ohio
DAVID L. MORRISON, Nuclear Regulatory Commission (retired), Cary, North Carolina
BRIJESH VYAS, Bell Laboratories-Lucent Technologies, Murray Hill, New Jersey
ROBERT D. WEAVER, Electric Power Research Institute (retired), Palo Alto, California

Liaison from the Board on Energy and Environmental Systems

ROY G. GORDON, NAS,[2] Harvard University, Cambridge, Massachusetts

Project Staff

JAMES ZUCCHETTO, director, Board on Energy and Environmental Systems (BEES), and study director (from April 1998)
JILL WILSON, senior program officer and study director (until April 1998)
PATRICIA SPAULDING, project assistant (until January 1998)
SUSANNA E. CLARENDON, financial and administrative assistant, and project assistant (from February 1998)

---

[1]NAE = National Academy of Engineering
[2]NAS = National Academy of Sciences

BOARD ON ENERGY AND ENVIRONMENTAL SYSTEMS

ROBERT L. HIRSCH (*chair*), Advanced Power and Energy Technology Collaborative, Inc., Washington, D.C.
RICHARD MESERVE (*vice chair*), Covington & Burling, Washington, D.C.
EVERETT H. BECKNER, Lockheed Martin Corporation, Albuquerque, New Mexico
JAN BEYEA, Consulting in the Public Interest, Lambertville, New Jersey
E. GAIL DE PLANQUE, NAE, Potomac, Maryland
WILLIAM L. FISHER, NAE, University of Texas at Austin
WILLIAM FULKERSON, Oak Ridge National Laboratories and University of Tennessee (retired), Knoxville
ROY G. GORDON, NAS, Harvard University, Cambridge, Massachusetts
EDWIN E. KINTNER, NAE, GPU Nuclear Corporation, Norwich, Vermont
K. ANNE STREET, Geo-Centers, Rockville, Maryland
JAMES SWEENEY, Stanford University, Stanford, California
LINDA GILLESPIE STUNTZ, Stuntz & Davis, Washington, D.C.
KATHLEEN C. TAYLOR, NAE, General Motors Corporation, Warren, Michigan
IRVIN WHITE, UTECH, Inc., Fairfax, Virginia

Liaisons from the Commission on Engineering and Technical Systems

RUTH M. DAVIS, NAE, Pymatuning Group, Inc., Alexandria, Virginia
LAWRENCE T. PAPAY, NAE, Bechtel Technology and Consulting, San Francisco, California

Staff

JAMES ZUCCHETTO, director
JILL WILSON, senior program officer (until April 1998)
TRACY WILSON, senior program officer (until April 1998)
SUSANNA E. CLARENDON, project assistant and administrative associate
PATRICIA SPAULDING, project assistant (until January 1998)

Acknowledgments

The committee wishes to thank the representatives of the United States Advanced Battery Consortium (USABC) and the U.S. Department of Energy who contributed significantly to this National Research Council (NRC) study, whether by giving presentations at meetings or providing responses to committee requests for information. The committee also acknowledges the valuable contributions of individuals from other organizations (see Appendix B). In addition, the committee is grateful for the able assistance of Dr. James Zucchetto, Dr. Jill Wilson, Ms. Susanna Clarendon, and Ms. Patricia Spaulding of the NRC staff for their hard work organizing and planning committee meetings and for their efforts in gathering information and writing sections of the report. Finally, the chair wishes to recognize the committee members for their expertise, their efforts, and their consistent commitment to integrity throughout the study.

This report has been reviewed by individuals chosen for their diverse perspectives and technical expertise, in accordance with procedures approved by the NRC's Report Review Committee. The purpose of this independent review is to provide candid and critical comments that will assist the authors and the NRC in making the published report as sound as possible and to ensure that the report meets institutional standards for objectivity, evidence, and responsiveness to the study charge. The content of the review comments and draft manuscript remain confidential to protect the integrity of the deliberative process. We wish to thank the follow individuals for their participation in the review of this report:

Jack L. Blumenthal, TRW Space and Defense Sector; Edwin Kintner, GPU Nuclear Corporation (retired); Donald MacArthur, CHEMAC

International Corporation; John Newman, University of California, Berkeley; Roberta Nichols, Ford Motor Company (retired); and Daniel Sperling, University of California, Davis.

While the individuals listed above have provided constructive comments and suggestions, responsibility for the final content of this report rests solely with the authoring committee and the NRC.

Contents

Tables and Figures

TABLES

FIGURE

Executive Summary

In January 1991, three major U.S. automakers—Chrysler Corporation, Ford Motor Company, and the General Motors Corporation (GM)—entered into an agreement to pool their technical knowledge and funding to accelerate progress in the development of batteries for electric vehicles (EVs). Their partnership, which is slated to run until 2003, is called the United States Advanced Battery Consortium (USABC). The purpose of the USABC is "to work with advanced battery developers and companies that will conduct research and development (R&D) on advanced batteries to provide increased range and improved performance for electric vehicles in the latter part of the 1990s." In addition to the three automakers, participants in the USABC include the U.S. Department of Energy (DOE), through a cooperative agreement with the partnership, and the Electric Power Research Institute (EPRI), through a participation agreement.

The USABC began its technical program by developing goals and objectives as a basis for selecting and managing projects. Midterm performance goals were designed to meet "introductory EV market requirements assuming EV subsidies." The long-term goals define a level of technology capable of sustaining a competitive EV market. In 1996, the USABC developed intermediate commercialization goals that specify technologies the USABC believes will allow the entry of EVs into the market without subsidies.

The USABC has awarded eight contracts to companies for work on battery technologies for EVs and is sponsoring related projects at several DOE national laboratories through cooperative research and development agreements (CRADAs). In Phase I (1991–1996), work was focused on six major battery systems: nickel metal hydride (Ni/MH), sodium-sulfur, lithium-ion, lithium-polymer, lithium-ion-polymer, and lithium-iron disulfide systems. Total spending for Phase I was

$189 million, which was divided equally between government (DOE) and industry (the three automakers, EPRI, electric utilities, and battery companies). In Phase II (1996–2000), the focus is on Ni/MH, lithium-ion, and lithium-polymer systems. Phase II funding is expected to total $106 million, of which DOE will provide 45 percent and industry 55 percent.

SCOPE AND CONTENTS OF THIS STUDY

This study by the National Research Council (NRC) was requested by DOE's Office of Advanced Automotive Technologies. The study focuses on the processes used by the USABC to select, evaluate, and manage R&D projects on EV batteries in Phases I and II of the program.

The chapters in this report are focused on the following topics: the management structure of the USABC; the establishment of performance and cost goals and the selection of projects; technical progress in Phases I and II; and the quality of management and oversight. The principal conclusions and recommendations presented below are grouped into five categories: overall perspective on the USABC; goals; program management; procurement; and battery technologies. A number of these recommendations are relevant to cooperative government-industry R&D programs in general, including the Advanced Battery Initiative, which has been proposed as a follow-on to USABC.

OVERALL PERSPECTIVE

Conclusion. The participants in the USABC have made a reasonable effort to develop technology for an extremely demanding application. Despite an overly ambitious schedule, which was imposed by regulatory requirements, the USABC has made significant progress. Although no technology has yet been brought to the point of fully meeting even the midterm goals, work in progress has a significant chance of meeting both the midterm and commercialization goals early in the next decade, possibly even before the year 2000. The consortium has effectively brought together battery technologists and experts in EVs in creative, focused partnerships but has missed some opportunities to provide broad technical leadership because timely information was not made public.

The USABC was created to respond to specific challenges imposed by regulatory agencies, and many aspects of its operation, including its technical strategy, have been affected by the regulatory timetable established by the California Air Resources Board (CARB). Technology capable of meeting the CARB mandate did not exist when the USABC was founded, and there was no obvious path to it. The USABC's experience has confirmed that there are major technical difficulties and challenges in developing batteries for EVs. Given the circumstances, the committee generally agrees with the strategic assumptions that have guided the USABC program, the most fundamental of which are (1) that development based on existing science should take precedence over research to create new science and (2) that multiple technologies should be pursued simultaneously.

The first two years of the USABC's existence were taken up with the logistics of creating the organization and its operating principles, defining goals, soliciting proposals, working out principles governing intellectual property, and negotiating contracts. In the committee's view, the USABC carried out these tasks as efficiently as could be expected, and many of the lessons learned by the USABC could be valuable for establishing future collaborative government-industry R&D programs. Nevertheless, the time remaining before the implementation of EVs in 1998, as required by the 1990 CARB mandate, was only about three years. The time required for development of the vehicle left very little time for development of the battery.

Nevertheless, the committee respects the role regulatory initiatives have played in forcing the pace of progress in this field, which can lead to significant public benefits as a result of reduced vehicle emissions. In the committee's view, if the implementation horizon had been after the year 2000, there would have been enough time for the USABC to pursue a more effective R&D portfolio for EV batteries and more opportunities for the USABC to pursue new science aimed at improving the performance and cost effectiveness of critical components. Regulators in this domain would be well advised to recognize that their time constraints have an impact on the range of science and technology that can realistically be brought to bear on a given problem.

By bringing together automotive manufacturers, battery manufacturers, electric utilities, and DOE, the USABC has provided a venue for a focused approach to meeting EV requirements. The interaction between automotive manufacturers and battery manufacturers has added a valuable dimension of systems integration and testing to the programs for battery development. Interaction among the partners on individual USABC projects has been very productive.

Nevertheless, the committee was surprised to find how little public information was being made available on an ongoing basis by the USABC, a multiyear program that receives substantial government funding. In the committee's view, the USABC could have fostered a climate in which battery technology could have progressed on a broader front and gone beyond large-scale efforts directed only toward relatively short-term applications. Because precompetitive issues, which could have been discussed in open venues, were not separated from private company issues, the USABC missed an opportunity for proactive leadership. The USABC partners clearly have a common long-term interest in fostering a healthy industry that has multiple suppliers for any given EV battery technology. The USABC could have promoted this interest effectively at very low cost by developing and updating a technology road map and by making concerted efforts to summarize progress regularly on a nonproprietary basis. More general information would also have helped to inform the public and policymakers on the difficulties of developing a commercially viable EV.

By adopting a collegial management model involving program officers from DOE, the USABC was able to take advantage of DOE's extensive expertise with battery systems. However, the committee is concerned that other DOE battery R&D (under the Exploratory Technology Research program or elsewhere in DOE) may have been unwisely reduced to pay for federal participation in the USABC, which is a focused development program aimed at one very demanding battery application that might be unrealizable in terms of the defined performance goals. The committee believes that the nation as a whole has many different interests in battery science and technology, and these interests should be managed, financially and otherwise, in a balanced way by DOE and other federal research programs.

A number of groups besides the USABC involving industry and government representatives are addressing the infrastructure required to support a significant EV market, such as standardized charging systems, the supplies of materials for the construction of advanced batteries, and battery disposal and recycling. The committee understands that the USABC considers these matters outside its scope, but they are critical to meeting USABC's goals. Therefore, the committee believes the USABC would be justified in paying more attention to infrastructure issues in the remainder of the program.

Recommendation. In future programs and current programs outside the USABC, DOE should focus more on research that advances the science of electrochemical systems that have the potential to meet the long-term performance and cost criteria for electric vehicles, as opposed to development of systems based on existing science. A significant portion of this scientific research should be focused on battery technologies that have not been included in the USABC program.

Recommendation. In concert with other groups addressing infrastructure issues, the USABC should use some of its remaining Phase II funds to address infrastructure for its most promising long-term technology, in particular, ensuring the availability of materials and recycling of batteries.

Recommendation. The USABC should communicate its targets and results on a regular basis to the technical community, within the bounds of agreements on proprietary information. The USABC should develop a technical road map to guide battery development even outside the USABC, and to identify barriers to the implementation of batteries in commercially viable electric vehicles. A technical road map would also be valuable for follow-on programs to the USABC.

Recommendation. Given the trend toward cooperative government-industry technology development and commercialization, DOE, should collaborate with other federal agencies to develop a governmentwide approach to the efficient implementation of cooperative agreements. These procedures could also be used for launching a follow-on program to the USABC.

PROGRAM GOALS

Conclusion. The USABC acted appropriately in establishing performance goals consistent with both the CARB mandate and the marketing assumptions made by the three major U.S. automakers. The goals reflect a legitimate view of the requirements for meeting the CARB mandate with vehicles that could be successful in the marketplace. A successor to the USABC may have the latitude to rethink the performance goals in the light of ongoing technical developments and changing market potential.

In terms of fundamental energy considerations, a battery-powered EV cannot match a gasoline-fueled internal combustion engine (ICE) vehicle[1] in range, performance, and recharging characteristics. However, many vehicles are used in circumstances where these characteristics are not critical. In establishing performance goals consistent with CARB requirements, the USABC assumed that EVs would be competitive only if their performance were comparable to the performance of ICE vehicles. The committee concluded that the USABC's approach was reasonable in light of the CARB regulatory requirements. The USABC relied heavily on its own marketing study and marketing studies by the U.S. automakers to compare the competitiveness of EVs and ICE vehicles. Marketing strategies based on the advantages of EVs, such as quietness, cleanliness, automatic in-home "refueling," and potentially reduced maintenance requirements, might have been used to establish more attainable performance goals.

The battery systems developed to meet the midterm goals were not expected to gain a sustainable share of the market for EV batteries or even to contribute to the development of battery systems that would meet the long-term goals. The committee observed that the main purpose of developing batteries to meet the midterm goals was to meet (or show a good faith effort at meeting) the original CARB mandate (which has since been changed) for a 2 percent new car market share in California for zero emmission vehicles (ZEVs) in the late 1990s. The committee questions the wisdom of using short-term regulatory requirements to establish an initial technical direction for the USABC's program.

Cost goals should be a high priority for the USABC program. The projected full cost of the ownership and operation of a battery is an important consideration in the economics of EV development. Although the USABC recognizes the importance of the cost issue, its cost estimates have not always been based on consistent methods and realistic assumptions.

Recommendation. Participants in a follow-on program to the USABC should allocate some program funds to examining a broader spectrum of electric vehicle concepts and related market opportunities, in addition to developing a replacement vehicle for a gasoline-fueled ICE vehicle that could capture a significant share of the conventional automobile market.

---

[1]In this report, an ICE vehicle is assumed to be a conventional, gasoline-fueled vehicle.

Recommendation. The participants in a follow-on program to the USABC should make a critical assessment of the economics of batteries for electric vehicles. The issue of cost should be addressed early in the program, and cost projections should be monitored and tested by consistent methods throughout battery development. Meeting cost goals should be an important criterion for maintaining funding.

PROGRAM MANAGEMENT

Conclusion. The three major U.S. automakers have managed the USABC program effectively using proven industry practices and have made appropriate decisions to narrow the portfolio of battery technologies as the program has proceeded.

In the committee's judgment, the industry-led consortium model adopted by the USABC has been effective in focusing development on well defined, but extremely challenging, technical targets for a specific application. Because the three major domestic automakers are customers for EV batteries themselves and are bound to ensure that batteries developed under the USABC are compatible with vehicle requirements, the committee considers it appropriate for the automakers to have overall responsibility for battery development. Moreover, the committee doubts that a purely government-managed program could have brought a similar focus or managerial crispness to the program, or that a purely private initiative could have brought the same level of resources to bear or engaged as wide a range of participants.

The committee observed that both funding and decision making within the consortium are highly leveraged. Although funding levels from the automakers and EPRI are proprietary, the committee estimates that the contribution from the automotive companies is significantly less than the contribution from DOE; nevertheless, most of the consortium's management decisions are made by the automakers.

In the committee's judgment, Chrysler, Ford, and GM have managed the program effectively and have made appropriate decisions to narrow the portfolio of battery technologies as the program has proceeded. Mechanisms have been established to ensure that the statements of work for each project are being followed or are modified in accordance with established procedures. Mechanisms have also been established for focusing R&D on critical issues, and informed Go/No Go decisions have been implemented.

Although the USABC program management has many excellent features, the committee believes that some changes in the internal peer review process would be beneficial. The USABC oversight management process does not include peer review as it is usually understood in the context of projects supported by government agencies. The USABC peer review process involves only personnel from the

USABC's management structure and, in the committee's opinion, lacks objectivity. Independent peer reviews by external experts at specified intervals—and at the start-up of major new projects—would enhance the USABC's management processes. The peer review process should include procedures for handling proprietary information and for communicating nonproprietary information to a broad technical audience. Input from a wide audience would increase the likelihood of the USABC finding the best midterm and long-term approaches to technological problems.

DOE has responsibilities both as a partner and as an overseer in the USABC. By virtue of the cooperative agreement between DOE and the USABC, DOE is substantially involved in approving all statements of work, selecting and approving all technologies, overseeing all major decisions, issuing technical directives, and approving the USABC's annual plan. On an operational basis, DOE's participation appears to the committee to be largely collegial rather than directive. Although this arrangement seems to be working smoothly and has had good results, it is probably not sound practice for individuals to play both collegial and oversight roles.

Recommendation. Within the limits imposed by proprietary considerations, regular peer reviews of the USABC's ongoing programs should be implemented immediately to provide objective assessments and to support decision making. Peer reviews should also be incorporated into any follow-on program that receives federal funding.

Recommendation. DOE should keep its collegial and oversight roles in the USABC program separate by ensuring that managers above the level of participants receive appropriate progress reports.

PROCUREMENT

Conclusion. The USABC has solicited proposals, chosen contractors, and made awards using accepted procurement processes. Although methods for verifying practical accountability have been established, the progress made by the USABC has not been satisfactorily disseminated to a broad audience. As a result, some opportunities for technical advancement have been lost. The USABC's policies on intellectual property and its processes for evaluating proposals may have precluded the participation of small companies and international partners, especially organizations with relatively mature battery systems.

The USABC's original request for proposal information (RFPI) was broad and open, and the solicitation process stimulated an impressive amount of creative collaboration among technology developers. For example, a joint project by 3M and Hydro-Québec, which was brokered by the USABC, appears to be a successful collaboration between owners of potentially valuable, complementary technologies. The results of this collaboration appear to be much greater than the sum of the parts.

In general, small companies were unable to compete for contracts effectively because the proposal evaluation criteria placed considerable emphasis on the developer having an established manufacturing capability. In addition, the USABC's fundamental objective of fostering American competitiveness in the automobile and battery industries tends to limit international collaborations. These comments are observations about, not criticisms of, the RFPI.

The USABC's policies governing intellectual property favor the partners of the consortium rather than the developers of the technology. Other models that the committee examined, such as the Sematech model, have adopted policies that are more favorable to the developers and, therefore, seem more likely to promote the enthusiastic development of a desired technology. In this respect, the announced practices in the RFPI appear to be defensive. The RFPI requirement that companies reveal commercially sensitive information almost certainly discouraged battery developers with relatively mature systems—which they have developed over many years at considerable expense—from participating.

The committee applauds the consortium's requirement that testable items, particularly cells, batteries, and packs, be delivered periodically. The committee also believes that the national laboratories have performed a valuable role as independent evaluators of technology, especially through testing programs tailored to the needs of the USABC. By this mechanism, the consortium has been able to develop an excellent method of establishing practical accountability. However, the virtual absence of timely progress summaries for a broader audience is a major deficiency. Proper documentation is not only a means of tracking progress, but is also a valuable tool for analyzing and stimulating researchers and vendors. Opportunities for technical advancement were probably lost because of the lack of timely documentation.

BATTERY TECHNOLOGIES

Conclusion. Given the constraints of the CARB mandate, the USABC made a reasonable selection of prospective technologies for EVs and pursued these technologies with appropriate partners. Substantial progress has been made even though none of the required technologies and none of the battery systems currently funded meets the USABC's cost or life objectives.

At the outset of the USABC program, the technology did not exist to support the program's goals. Although those technologies have still not been fully developed, substantial progress has been made in the development of Ni/MH and lithium battery technologies for EV applications, as well as in design, manufacturing processes, system integration, and cost reduction.

In the committee's judgment, the USABC acted appropriately in closing down unpromising lines of inquiry in a timely manner, especially because the USABC had to focus on relatively mature systems that offered some chance of meeting the program goals within a very short time frame. The USABC was correct to cancel the sodium-sulfur, lithium-iron disulfide, and lithium-ion-polymer battery projects, which did not meet critical performance criteria or offer promising business opportunities by the end of Phase I. The committee also considers that the USABC acted appropriately in focusing Phase II on reducing battery cost. None of the battery systems presently funded meets its cost objectives, which will be critical to the successful commercialization of EVs. Commercialization will also depend on battery cycle life and calendar life, which have yet to be determined for most systems, and on satisfactory battery performance over a range of operating conditions.

The midterm technology programs were established in response to the CARB goal of requiring 2 percent of new cars in California to be ZEVs in 1998. However, CARB has changed the mandate to 10 percent in 2003 and is considering revising that mandate to include hybrid vehicles. The committee, therefore, believes that the USABC is justified in reevaluating its midterm battery programs. The Ni/MH battery, for example, which does not meet the midterm performance or cost requirements, will be discontinued in 1998 although it will remain under development in a separate program to meet the high power energy storage requirements for hybrid vehicles for the Partnership for a New Generation of Vehicles (PNGV) program. The lithium-ion battery, which meets the midterm performance criteria but not the cost requirements, may be worth pursuing if it has the potential to meet the interim commercialization criteria.

Recommendation. In Phase II, the USABC should place more emphasis on evaluating promising battery systems under extreme temperatures and stressful-use conditions. Batteries should also be subjected to extensive safety and reliability testing from the beginning to the end of life.

Recommendation. The USABC should reevaluate its midterm battery programs. The development of the lithium-ion battery should be continued if it has the potential to meet the interim commercialization criteria.

1

Purpose, Scope, and Process of This Study

In September 1990, the California Air Resources Board (CARB),[1] an agency of the California Environmental Protection Agency, adopted regulations for low-emission vehicles and clean fuels as part of its continuing efforts to reduce pollution from motor vehicles. These regulations established stringent standards for tailpipe emissions for new classes of vehicles and defined a yearly average fleet emission rate, beginning in 1994. The most stringent emissions standards applied to "zero-emission vehicles" (ZEVs), which produce no smog-forming tailpipe or evaporative emissions. According to the 1990 CARB mandate, ZEVs were scheduled to comprise 2 percent of the new vehicles sold in California by every large automobile manufacturer,[2] starting in 1998. The percentage of ZEVs was scheduled to increase to 5 percent in 2001 and to 10 percent in 2003. In 1996, the intermediate requirements for 1998 and 2001 were dropped, and the current requirement is that 10 percent of new vehicles must be ZEVs by 2003. Massachusetts has a similar mandate, and New York is trying to establish a 2 percent ZEV requirement.

ELECTRIC VEHICLES

The only vehicle currently capable of meeting the California ZEV requirements is the battery-powered electric vehicle (EV). From the perspective of the

---

[1]CARB was founded in 1967 "to promote and protect public health, welfare, and ecological resources through the effective and efficient reduction of air pollutants, while recognizing and considering the effects on the economy of the state" (CARB, 1967). The 11 members of CARB are appointed by the governor of California.

[2]The mandate applied to seven car companies with sales of more than 35,000 vehicles per year in California: General Motors, Ford, Chrysler, Toyota, Nissan, Honda, and Mazda.

complete fuel cycle, however, battery-powered EVs do not have zero emissions. For example, the power plant produces emissions associated with the production of electricity for recharging the vehicle. However, CARB argues that it is easier to control emissions from a small number of stationary sources (power plants) than from millions of mobile sources (automobiles) (CARB, 1996).[3]

Chrysler Corporation, Ford Motor Company, and the General Motors Corporation (GM) anticipate that customer requirements for a sustainable EV market will include a vehicle range of significantly more than 100 miles in actual customer use; life-cycle costs comparable to those of gasoline-fueled internal combustion engine (ICE) vehicles;[4] performance equivalent to the performance of an ICE vehicle, including hill climbing ability; minimal routine maintenance; and the ability to operate in all climates (R. Davis, 1997).

The batteries that can be accommodated in an EV equivalent in size to a typical family sedan cannot provide as much energy as the gasoline in the fuel tank of a conventional vehicle. Consequently, the range of an EV on a single battery charge is far less than the 300-mile range per tank of fuel for a typical gasoline-fueled vehicle with an ICE. An EV with a conventional chassis and body that uses off-the-shelf drivetrain components and a conventional lead-acid battery has a practical range of about 50 miles (CARB, 1994). CARB reports that the range can be increased to about 100 miles by incorporating recent advances in motor and electronic technology. The incorporation of energy efficient vehicle technologies, such as lightweight materials, regenerative braking systems, and improved motors and controllers, can further increase the range. However, the development of improved batteries remains the biggest technological challenge for EVs.

The cost of existing batteries is also a barrier to the widespread consumer acceptance of EVs because the initial purchase price of an EV is almost certain to be higher than the price of a gasoline-fueled vehicle. Whether or not advances in technology and mass production can reduce the life cycle costs of an EV to those of a gasoline-fueled vehicle is an open question.[5]

U.S. ADVANCED BATTERY CONSORTIUM

The CARB mandate for the introduction of ZEVs created a resurgence of interest in EVs and related technologies. In January 1991, the three major U.S. automakers—Chrysler, Ford, and GM—entered into an agreement to pool their

---

[3]CARB is considering an amendment to its requirement for ZEVs that would allow any vehicle to qualify for ZEV credit if its emissions are no higher than the power plant emissions associated with an EV and its emissions do not increase over time (in contrast to those of a conventional vehicle). Hybrid electric vehicles and fuel cell-powered vehicles may be able to meet these criteria.

[4]In this report, an ICE vehicle is assumed to be a conventional gasoline-fueled vehicle.

[5]For further information on EV battery technology, see the study sponsored by CARB of EV battery technology from an international perspective (Kalhammer et al., 1995).

technical knowledge and funding in an effort to accelerate progress in the development of batteries for EVs. Their partnership, which is slated to run until 2003, is called the United States Advanced Battery Consortium (USABC). The purpose of USABC is to "work with advanced battery developers and companies that will conduct research and development (R&D) on advanced batteries to provide increased range and improved performance for electric vehicles in the latter part of the 1990s" (R. Davis, 1997). In addition to the three automakers, participants in USABC include the U.S. Department of Energy (DOE), through a cooperative agreement with the partnership, and the Electric Power Research Institute (EPRI), through a participation agreement.

The USABC has awarded a number of contracts to battery companies for work on battery technologies for EVs and has sponsored related projects at several DOE national laboratories, through cooperative research and development agreements (CRADAs). Phase I, from 1991 to 1996, focused on six major battery systems: nickel metal hydride (Ni/MH), sodium-sulfur, lithium-ion, lithium-polymer, lithium-ion polymer, and lithium-iron disulfide systems. Total spending for Phase I was $189 million, which was borne equally by government (DOE) and industry (the three domestic automakers, EPRI, electric utilities, and battery companies). Phase II, from 1996 to 2000, will focus on Ni/MH, lithium-ion, and lithium-polymer systems. Total Phase II funding is expected to be $106 million, 45 percent of which will be provided by DOE and 55 percent of which will be provided by industry.

ORIGIN AND SCOPE

At the request of the director of DOE's Office of Advanced Automotive Technologies (OAAT),[6] the National Research Council (NRC) convened a committee under the auspices of the Board on Energy and Environmental Systems to conduct a retrospective examination of the processes used by the USABC to select and evaluate R&D projects on EV battery technology in Phase I of the program. The committee was also asked to comment on the USABC's plans for the selection of Phase II projects. Some of the topics the committee was requested to address are listed below:

- the process by which technical goals and objectives were established for EV battery development
- the process used by the USABC to solicit proposals, choose contractors, and make awards, both for new projects and for continuing efforts
- the manner in which contractor performance was measured and evaluated by the USABC

[6]DOE funding for USABC activities is provided through OAAT.

- how actual contractor results measured up against the technical goals and objectives
- the USABC's plans for Phase II[7]

Thus, the scope of the committee's task was oriented towards processes used in the USABC program. The goals of the program were established to fulfill the CARB mandate and the automotive companies' analyses of the market for EVs. The committee's review is based on the established goals, even though, in the committee's opinion, 1998 was not a realistic date.

Committee members included experts in battery technology and electric power systems, R&D management, materials science, electrochemistry, and automotive applications of advanced technologies. Biographical sketches are provided in Appendix A.

PROCESS AND ORGANIZATION

The committee met twice between October and December 1997. Most of the first meeting was devoted to presentations by USABC personnel, who described the processes used to establish technical goals and objectives for battery development; to solicit proposals, choose contractors, and make awards; and to manage projects and evaluate contractor performance (see Appendix B). At the second meeting, additional presentations were made on lithium battery technologies and government-industry R&D consortia. Most of the second meeting was devoted to committee deliberations. In addition to the presentations at meetings, the committee received written information from the USABC on the project selection process and on technical progress (and problems) in Phases I and II. One committee member visited the 3M facility in Minneapolis where lithium-polymer battery facilities have been developed by 3M and Hydro-Québec under the USABC program. Some of the material, called "USABC protected battery information," provided to the committee under an agreement signed by the National Academy of Sciences, the USABC, and DOE, included proprietary information of the USABC or its member companies.

General information on advanced battery development was obtained by individual committee members in the course of discussions with members of the broad technical community. These discussions also provided insights into the overall impact of the USABC program on battery R&D in the United States.

Chapter 2 of this report provides background information on the structure and funding of the USABC and summarizes the roles of the industry participants

---

[7]When the proposal for this study was submitted to DOE, Phase II was in the planning stage. However, by the time the study got under way, the major Phase II contracts had been awarded. The committee was, therefore, able to comment on both the plans for Phase II and progress through the end of calendar year 1997.

and the government in the consortium's activities. Chapter 3 provides a critique of the technical goals and objectives established by the USABC for battery development, as well as the committee's assessment of the processes used to solicit and select R&D projects. Chapter 4 contains the committee's evaluation of technical strategies and progress in Phases I and II. Chapter 5 contains the committee's evaluation of the USABC as a management model—including comparisons with other government-industry R&D partnerships—and an assessment of the USABC's oversight of R&D projects.

Chapter 6 provides the committee's conclusions and recommendations, based on the findings in Chapters 3 through 5. The recommendations address specific features of the USABC program, especially activities scheduled between now and the termination of the program in 2003, as well as generic issues relevant to government-industry R&D partnerships. In developing its conclusions and recommendations, the committee was mindful of topics identified by the OAAT director as being of particular interest to DOE (Patil, 1997). These topics include the selection of lithium-polymer battery technology as the long-term option for EVs; the balance between research and development in the USABC program; opportunities for the national laboratories, small businesses, and universities to participate in the USABC program; the management of government-industry R&D partnerships; and the possibility of a follow-on program to the USABC after 2000.

2

Nature and Structure of the U.S. Advanced Battery Consortium

The three biggest domestic automakers, Chrysler, Ford, and GM, established the USABC in early 1991 to accelerate the development of batteries for EVs. Because EVs could help reduce mobile-source air pollution and allow electric utility companies to use their excess capacity during off-peak hours, EPRI, along with several individual utility companies, agreed to participate in the consortium in mid-1991. EPRI had funded the development of batteries independently for more than 15 years. Responding to a legislative mandate to pursue the benefits of EVs,[1] DOE agreed to cooperate with the consortium in late 1991. The relationship between DOE and the three automotive manufacturers is governed by a cooperative agreement that requires DOE to be actively and substantially involved in the management of the consortium (DOE, 1991).

The USABC is one of about a dozen partnerships among Chrysler, Ford, and GM, who have joined forces under the U.S. Council for Automotive Research (USCAR) to research and develop precompetitive automotive technologies. The USABC is one of the largest ventures under the USCAR agreement and is the only USCAR consortium that currently receives significant federal funding (Thorpe, 1997).

---

[1]The legislative mandate is contained in the Electric and Hybrid Vehicle Research, Development, and Demonstration Act of 1976. The Energy Policy Act of 1992 (EPACT) reaffirmed this mandate and authorized the secretary of energy to enter into cooperative agreements with industry to develop advanced batteries for EVs.

FUNDING

Total funding for Phase I of USABC was originally planned to be $262 million, 50 percent of which was to be provided by DOE and 50 percent by industry (i.e., the three automotive companies, battery developers, and the utility industry [GAO, 1995]). In fact, the total spent during Phase I was approximately $189 million (Heitner, 1998; Thorpe, 1997). The shortfall was partly attributable to the USABC's decision to terminate some projects early because of their lack of progress toward meeting objectives. In addition, some projects were delayed because of the logistics of setting up the consortium and the time required to establish operating principles and procedures.

DOE and industry contributed equally to Phase I funding, as originally planned. Under the cooperative agreement with DOE, the three USABC automotive partners were responsible for the entire nongovernmental half of the budget. (Cost-sharing provisions in the USABC contracts provide for funding from outside the three automotive companies.) The participation agreement between EPRI and the USABC ensures contributions from the utility industry. Battery developers also contributed on a negotiated basis. Representatives of the USABC told the committee that when the Phase I contracts were negotiated, the consortium hoped that between 30 and 50 percent of the costs would be paid by battery developers (Thorpe, 1997). Detailed information on the financial contributions of the automakers and EPRI is proprietary. However, the committee noted that the financial contribution from Chrysler, Ford, and GM, which is leveraged not only by DOE, but also by other industry participants in the USABC, is significantly less than the financial contribution of the federal government.

Some of the unspent DOE funding for Phase I is being carried forward to Phase II. The total budget for Phase II is anticipated to be $106 million, which will again be split between government and industry, with DOE providing 45 percent and industry 55 percent.

Total DOE funding for R&D on EV batteries for fiscal years (FYs) 1990 through 1998 is shown in Table 2-1. In FYs 1990, 1991, and 1992, when DOE's Office of Energy Efficiency and Renewable Energy was reorganized and the USABC were formed, DOE began to phase out its existing development programs on advanced batteries, which were replaced by the USABC projects. DOE also reduced its long-term high-risk work under the Exploratory Technology Research (ETR) program, which seeks to identify new rechargeable battery systems with higher performance and lower life-cycle cost than those of current systems and to provide critical supporting and materials research for batteries under development by the USABC. The overall goal of the ETR program is to develop electrochemical power sources suitable for use in electric and hybrid vehicles.

TABLE 2-1 DOE Funding for Advanced Battery Research (millions of dollars), FYs 1990 to 1998

FY	DOE Development Programs and Supporting Work	Exploratory Technology Research Program	USABC Cooperative Programs (Phase)[b]
1990	7.1	5.6	—
1991	4.1	4.6	6.4 (I)
1992	0.5	2.6	21.0 (I)
1993	2.5[a]	3.9	21.9 (I)
1994	0.3	3.3	26.9 (I)
1995	0.2	2.1	22.2 (I)
1996	0.4	1.9	15.0 (II)
1997	0.0	2.4	13.0 (II)
1998 (planned)	0.3	3.5	12.2 (II)

[a]Funding for the development of an air battery that was not part of the USABC program.

[b]These funds include expenditures for CRADAs with several of the national laboratories of about $32 million (GAO, 1995). Work included development and testing for Ni/MH, sodium-sulfur, lithium-iron disulfide, and lithium-polymer battery technologies. As far as the committee is aware, little, if any, of these funds was used to support new science for possible application to advanced battery systems.

Source: Sutula, 1997.

MANAGEMENT

The USABC is governed by a partners' committee of senior executives from Chrysler, Ford, and GM; the partners' committee sets USABC policy. A management committee made up of managers from Chrysler, Ford, and GM makes business decisions, selects contractors, and determines funding, in concert with representatives of DOE and EPRI. The posts of chair, treasurer, external affairs officer, and secretary are rotated annually among Chrysler, Ford, and GM. In Phase I, representatives of EPRI were nonvoting members of the management committee. In Phase II, however, EPRI has a voting member. Another EPRI representative managed the GM/Ovonic project. A DOE representative contributes to business decisions, participating in the selection of contractors and the determination of funding for projects. All decisions are based on a consensus management process.

The responsibility for executing technical programs is vested in the technical advisory committee (TAC), which includes the program managers for each USABC-funded battery technology, as well as technical representatives of the three automakers, EPRI, the electric utilities, the national laboratories, and DOE managers. The USABC business manager, a representative of CARB, and the

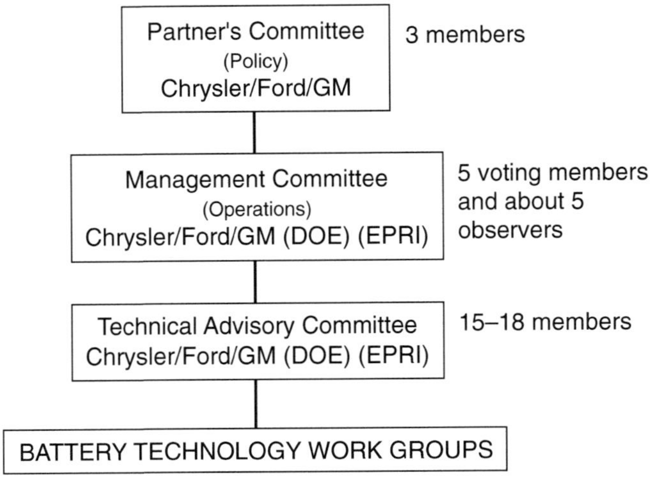

FIGURE 2-1 USABC management structure. Source: Thorpe, 1997.

program managers from DOE serve on the TAC. TAC program managers are supported by program teams, or work groups, for each technology; members of the teams are drawn from the TAC (see Figure 2-1). These teams evaluate proposals, negotiate contracts, and manage USABC technical programs with battery companies and developers. The fields of expertise of team members include vehicle integration, manufacturing, problem solving, testing, diagnostic analysis, and basic chemistry. The TAC also manages relationships with DOE's national laboratories that conduct USABC-funded research and testing.

The function of USABC program managers is similar to the function of program managers in industry and governmental agencies responsible for programs that involve many internal and external components. The USABC program managers, who are selected for their technical knowledge and management skills, are all employees of the three automakers or EPRI and its participating utilities. Most of them have been trained in management by their respective companies, and the USABC provides some additional program management training. One program manager is assigned to lead each USABC battery development project.

Historically, program managers have been assigned to one battery development program at a time, although recently one individual managed both a high-power battery contract and an EV battery contract that involved the same battery company. Overlapping responsibilities are permitted only if there are no conflicts of proprietary information. Program managers typically review projects every two to four weeks, accompanied by experts from the TAC, when necessary. Quarterly in-depth reviews are conducted by the entire work group, including the

DOE manager. These reviews involve visits to the developers that last a day or a day-and-a-half. The work group assesses the status of all tasks, goals, and objectives of the program, does an in-depth review of issues, and discusses future directions. Changes in the timing or objective of an individual task are based on a consensus of the work group.

Higher level peer reviews and critical reviews performed by members of the TAC are used for reporting the project status and issues to the management committee, which makes decisions about future activities. Most peer reviews involve all members of the TAC, including the managers from DOE, except when program managers may not be allowed to review proprietary information from competing programs. The program manager for the GM/Ovonic Ni/MH (nickel metal hydride) program, for example, does not have access to proprietary information relating to the SAFT Ni/MH program, and vice versa. These reviews, called critical reviews, are similar to peer reviews but are conducted by an appropriate subgroup of the TAC.

ROLE OF THE U.S. DEPARTMENT OF ENERGY

DOE's Office of Transportation Technologies (OTT), through the OAAT (Office of Advanced Automotive Technologies), provides management, contributes technical input, and oversees the USABC's battery development projects, as defined in a cooperative agreement between DOE and the USABC. Other DOE headquarters staff deal with legal and contractual issues pertaining to the consortium. DOE's Chicago Operations Office and various other offices are also involved in supporting DOE activities with the USABC. Within OTT, support is provided by the Office of Technology Utilization Field Operations Program, which tests and evaluates vehicles with advanced batteries as they become available.

Most of the R&D on advanced batteries has been conducted under competitive contracts awarded by the USABC to various organizations. However, some projects are carried out through CRADAs with five of DOE's national laboratories. For example, CRADAs were established with some of the national laboratories to develop test procedures, conduct independent tests and evaluations of batteries developed in the USABC program, and pursue R&D in each of the six battery programs. In addition, long-term, high-risk/high-payoff exploratory work is being conducted directly by DOE under OAAT's ETR program. Coordinating the activities of the USABC and federal advanced battery programs requires high-level programmatic direction, major program reviews, workshops, meetings, and symposia with experts on topics related to advanced EV batteries. Interagency coordination of DOE and the Defense Advanced Projects Research Agency projects is based on a memorandum of understanding. The government-sponsored Interagency Advanced Power Group is responsible for coordinating other projects and programs.

INFRASTRUCTURE ISSUES

Achieving a significant market penetration for EVs will require that a number of infrastructure needs, as well as environment, health and safety issues, be addressed. For example, standards for recharging vehicles, the availability of materials, and recycling of batteries are important infrastructure areas. While the USABC has been focused on battery development, a number of other groups have been working on broader issues. An Infrastructure Working Council was established in the early 1990s by EPRI, major electric utilities, and the automakers and is supported by DOE. DOE has also sponsored other activities to encourage the creation of the needed infrastructure for EVs. DOE assigned programmatic responsibility for environmental, health, and safety issues associated with advanced batteries, including recycling, to the National Renewable Energy Laboratory, which has conducted assessments of the advanced batteries under development by the USABC. The National Renewable Energy Laboratory has led the Advanced Battery Readiness ad hoc Working Group, which provides a forum for information for technology developers, the USABC, and regulatory agencies.

3

Performance Goals and Project Selection

The USABC began by developing objectives and goals at several levels. The goals and objectives were then used as the basis for the selection and management of projects. The objectives and goals, and their underlying assumptions and implications, are described below.

PROGRAM OBJECTIVES

The USABC established four overarching objectives (Thorpe, 1997):

- to establish a capability for an advanced battery manufacturing industry in the United States
- to accelerate the market potential of EVs through joint research on the most promising advanced battery alternatives
- to develop electrical energy systems capable of providing EVs with ranges and performance levels competitive with petroleum-based vehicles
- to leverage external funding for high-risk, high-cost R&D on advanced batteries for EVs

Operational targets based on the CARB mandates were (1) the development of a midterm battery, with production of a pilot-plant prototype in 1994 and (2) the demonstration of a feasible design for a long-term battery in 1994. The USABC's initial intention was to advance existing development-level projects to pilot-plant status. Once in pilot-plant production, advanced batteries could begin to be introduced into the EV market. Leveraging was achieved by cost sharing (roughly 50/50) between DOE and industry. Industry costs were typically split among the three automakers and the USABC contractors. If industry decided to produce and commercialize advanced batteries, they would bear all the costs.

ASSUMPTIONS ABOUT MARKET REQUIREMENTS

No known chemistry can be integrated with a practical battery structure that can store enough energy per unit volume to power an EV with the performance and range of an ICE vehicle. Substituting batteries for a gas tank on a volume-for-volume basis would be even more difficult. In the literal sense, the USABC's goal of developing an electrical energy system, excluding fuel cells, with range and performance levels that are *fully competitive* with ICE vehicles is doomed to failure on the basis of energy content and weight penalties. Preserving performance levels requires sacrificing vehicle range. The critical question is whether EVs will have a long enough range so that users do not feel restricted.

The USABC marketing data showed that 60 percent of new retail buyers desired a range of 100 to 125 miles under full load, with services such as heater, air conditioner, and lights (R. Davis, 1997). Only 30 percent of the buyers were interested in a vehicle with a range of 75 to 100 miles. The 125-mile range was based on a "need" of 50 miles per day, a "want" of 50 miles per day for contingencies, and the equivalent of 25 miles per day for accessories (e.g., a heater or air conditioner). The analysis did not provide an estimate of the percentage of customers that would actually purchase an EV. Also, the USABC data did not indicate the market share that could sustain the business.

The USABC market analysis revealed that customers would accept only a modest price increase (no more than a few thousand dollars) for an EV in comparison to an equivalent gasoline-fueled ICE vehicle. Thus the EV market is price-sensitive, which was borne out in the poor showing of lease-marketed vehicles in California in 1997. For example, GM has leased only a few hundred vehicles in Los Angeles, San Diego, Tucson, and Phoenix combined (Leclair, 1997.)

On the basis of the market analysis and experience with vehicle markets, the USABC set seven customer requirements for a sustainable market:

- reliable and safe operation under both routine and extreme conditions
- a range of more than 100 miles in actual customer use
- life-cycle costs comparable to those of an ICE vehicle
- equivalent performance in hill climbing to an ICE vehicle
- battery lifetime equal to the lifetime of the vehicle
- minimal routine maintenance
- operating capability in all climates

When the USABC was established, these market objectives could not be met by an EV powered by a commercially available battery. Furthermore, no extant battery technology had the obvious potential of meeting all of the customer and USABC expectations.

The customer use and acceptance defined by CARB as a basis for its quantitative ZEV goals were somewhat different from those defined by the USABC.

The CARB assumptions were more optimistic and were focused on a particular market niche. In addition to the environmental benefits of ZEVs, CARB identified numerous other benefits, such as daily home refueling, reduced emissions testing, and quiet performance.

CARB targeted a segment of the market where the vehicle would not have to compete with the ICE vehicle in all functions. CARB anticipated that most new car buyers would have access to more than one vehicle (e.g., the two-car family) and that for one of the vehicles the miles-per-day requirement and the need to take long trips would be greatly reduced or eliminated. A range of 100 miles per day would mean the EV could be driven more than 30,000 miles per year. CARB did not specifically account for the extra load on the battery for accessories.

PERFORMANCE GOALS

Throughout most of Phase I, the USABC's development strategy was based on two levels of battery performance, "midterm goals" and "long-term goals," respectively (Table 3-1). The midterm goals were designed to meet "introductory EV market requirements assuming EV subsidies." This level of performance would not support a sustainable market in terms of the seven customer requirements listed above. The long-term battery goals define a level of technology capable of sustaining a competitive EV market. The midterm performance levels were originally regarded as challenging but attainable targets that would enable automakers to respond to the CARB mandates for the late 1990s. Technical accomplishments in pursuit of the midterm goals were not necessarily expected to be useful for meeting the long-term goals. In addition to the performance goals given in Table 3-1, secondary criteria relating to characteristics, such as self-discharge, maintenance, and abuse resistance, were also established.

The USABC developed a third set of performance goals for Phase II, so-called commercialization goals, which are more demanding than the midterm goals but less ambitious than the long-term goals. Commercialization goals (also shown in Table 3-1) define conditions that the USABC judged would allow the entry of EVs into the commercial market without subsidies. The commercialization goals have been the principal focus of the program since 1996.

OPERATING COSTS

The USABC battery performance goals, which have played a major role in the USABC's decision making, are based on an analysis conducted in 1994 comparing the operating costs of EVs and ICE vehicles (R. Davis, 1997). Table 3-2 shows the performance values assumed in the USABC analysis; it was also assumed that the cost of the electric propulsion motors and supporting electronics for an EV would be the same as the cost of the engine, transmission, and supporting systems for an ICE vehicle. No data supporting this assumption were presented to the committee. Table 3-3 shows the calculated costs.

TABLE 3-1 USABC Performance Goals

	Midterm Goals	Commercialization Goals	Long-Term Goals
Specific energy (Wh/kg)	80 (100 desired)	150	200
Energy density (Wh/L)	135	230	300
Specific power (W/kg)	150 (200 desired)	300	400
Power density (W/L)	250	460	600
Life (years)	5	10	10
Cycle life (cycles)	600	1,000	1,000
Ultimate price ($/kWh)	< 150	< 150	< 100
Operating environment	– 30 to 65°C		– 40 to 85°C
Recharge time (hours)	< 6		< 3 to 6
Continuous discharge in 1 h[a]	75%		75%
Power and capacity degradation[b]	20%		20%

[a]Continuous discharge capacity is defined as the energy delivered in a *constant power discharge* required by an EV for hill climbing and high-speed cruising, specified as the percent of energy capacity delivered in a one-hour constant power discharge.

[b]Performance degradation defines the extent to which the battery system is unable to meet the original performance specification. The end of battery life corresponds to either a 20 percent reduction in acceleration power at 80 percent depth of discharge or a 20 percent loss of energy capacity.

Source: R. Davis. 1997; DOE, 1996.

The USABC's analysis shows that technologies that can truly compete on a cost-per-mile basis with an ICE vehicle will have to meet the long-term performance goals. But the analysis is very sensitive to the underlying assumptions. For example, the midterm goal for total battery cost, which is really "battery price," is lower by a factor of three to five than the price that was realized in the Phase I program. If the more realistic number of $600/kWh were used, the total midterm cost would be $0.25/mile, which is much higher than the cost with other batteries. Different assumptions about infrastructure and power costs would have changed

TABLE 3-2 Performance Assumptions Made by the USABC in Comparative Cost Analysis

	Lead-Acid Battery	Ni-Cd Battery	Midterm Goals	Long-term Goals	ICE Vehicle
Range (miles)	50	60	100	150	300
Battery cost ($/kWh)	150	500	150	100	na
Total battery cost ($)	1,875	7,500	3,750	3,750	na
Battery life (years)	2.5	5	5	10	na
Miles/year	10,000	10,000	12,500	15,000	15,000

Source: R. Davis, 1997.

TABLE 3-3 Calculated Costs from the USABC Analysis

	Lead-Acid Battery	Ni-Cd Battery	Midterm Goal	Long-Term Goal	ICE Vehicle
Battery cost/yr ($)	750	1,500	750	375	0
Battery cost/mi ($)	0.08	0.15	0.06	0.03	0
Energy cost/mi ($)	0.01	0.01	0.01	0.01	0.04
Total cost/mi ($)	0.09	0.16	0.07	0.04	0.04

Source: R. Davis, 1997.

the total cost per mile significantly. Assumptions about battery life also affected the calculated costs. The assumptions anticipate dramatic cost and performance advances in some batteries but not others. For example, if lead-acid battery life exceeds 2.5 years (as some evidence suggests), the calculated cost per mile given in Table 3-3 would be significantly lower. Other assumptions, such as driving 50 miles per day for 200 working days per year, seem to be consistent with commuter expectations.

SELECTION OF PROJECTS

The committee was asked to comment on the processes used by the USABC to solicit proposals, choose contractors, and make awards, for both new and continuing projects. The committee attempted to determine whether the USABC's procurement procedures were designed to select organizations with the technology (including manufacturing capabilities) and motivation most likely to result in batteries that meet USABC goals. Despite substantial technical progress (see Chapter 4), the USABC has failed to meet all of its midterm or long-term goals. Thus, the committee considered it important to determine whether the procurement process was responsible in any way for the lack of overall success of the USABC program.

Request for Proposal Information

The USABC's procedures for selecting Phase I collaborative R&D projects involved public announcements of the consortium's initiative, the preparation and release of a request for proposal information (RFPI), and the evaluation of proposals. The USABC mailed out 130 RFPI packages and received 59 proposals based on 16 different battery technologies (Nichols, 1992). Eight awards were made for six technologies.[1]

[1]The committee did not attempt to determine if the contract awards were consistent with the selection criteria used to evaluate proposals.

The RFPI is a detailed document that outlines the purpose and objectives of the USABC and reviews the prospective market for EVs. Primary and secondary technical criteria for midterm and long-term batteries are included, and two major program goals are highlighted: (1) the demonstration of production capability for the midterm battery in 1994 and (2) the demonstration of design feasibility for the long-term battery in 1994.

The RFPI spells out the expectations of the USABC for potential participants, including integrated team efforts to address well defined milestones and rapid progress toward prototypes because of the very short time scales set by CARB for the introduction of ZEVs in 1998. The specification of deliverables to be tested independently by DOE's national laboratories showed that this was an engineering, rather than a science, initiative. After issuing the RFPI, the USABC management urged interested parties to form partnerships that would complement their individual capabilities, a strategy the committee endorses.

Companies responding to the RFPI were required to provide information on company background; the proposed advanced battery technology; the technology development plan; schedule, deliverables, cost, and cost sharing; and cooperative arrangements. The requirement that participants share in the costs of the program all but eliminated contractors who did not have realistic expectations of achieving their goals.

The 59 proposals received in response to the RFPI were evaluated according to selection criteria based on the potential of the proposed technology to meet performance goals; the vision of the proposed development plan; the ownership of the proposed technology; the capability for high-volume manufacturing; cost sharing; the financial viability of the company/team; experience and past performance; the probability of successful commercialization; the qualifications of key personnel; the understanding of the project requirements; previous experience with high-power batteries; the ability to comply with battery packaging constraints; and the ability to meet required time schedules. Weighting factors were assigned to each category, with some values varying between midterm and long-term goals.

In the committee's judgment, some features of the RFPI might have put some potential contractors at an unnecessary disadvantage or discouraged the participation of some organizations. These features include the absence of specific guidelines for cost sharing; an apparent lack of incentives to develop midterm batteries; the requirement that technology be in hand; and the agreements for proprietary information and intellectual property. Nevertheless, the USABC's selection of projects was not biased. In the committee's opinion, the extensive knowledge of the status of battery technologies provided by DOE and other consortium participants was an important factor in the selection of appropriate proposals.

As Phase I proceeded, some technologies that were not sufficiently developed were eliminated by the consortium, generally by mutual agreement with the developers, and selected development projects were discontinued by individual companies (see Chapter 4). When Phase I ended and Phase II began, the only remaining candidate that could meet long-term goals was lithium polymer battery technology. To stimulate alternatives and possible competing technologies, another RFPI was issued but did not elicit any responses the USABC decided were worth funding.

Domestic and International Participation

The original RFPI was sent to all known U.S. and European companies that were working on or had developed battery technology. Asian companies were not invited to participate in Phase I because an Asian battery consortium was already well established (Stroven, 1997), and Asian companies already have major science, technology, and manufacturing capabilities for batteries. The USABC's decision to exclude them appears to be at odds with two of the objectives of the USABC initiative, namely, (1) accelerating the market potential of EVs through joint research on the most promising battery technologies and (2) providing electrical energy systems capable of providing EVs with range and performance levels competitive with those of petroleum-based vehicles. But including Asian companies could have conflicted with another USABC objective, establishing a capability for a United States advanced battery manufacturing industry. Thus, the exclusion of Asian participants from Phase I reflects the USABC's higher priority on fostering U.S. competitiveness than on the benefits of Asian advances in battery technology.

The committee noted that the USABC did not exclude European manufacturers in Phase I, which appears to be inconsistent with its treatment of potential Asian partners. SAFT America, Inc. (the American arm of the French company, SAFT), VARTA (a German company) and the American office of Silent Power (which was owned by the German RWE group in 1993) all participated (see Chapter 4). However, it was not clear to the committee that members of the Asian battery consortium would have been interested in working within the USABC framework because their program was already well established, and they would probably not have been willing to reveal commercially sensitive information as required by the RFPI (see below). No responses to the Phase II RFPI were received from Asian companies, even though the USABC solicited their participation in an effort to diversify its portfolio of long-term battery technologies.

Battery Technologies

The responses to the Phase I RFPI addressed 16 battery technologies, although the lead-acid system was deliberately excluded from consideration, based, in part, on the fact that Chrysler, Ford, and GM were already working individually with lead-acid battery manufacturers in a competitive environment.[2] Collaborative USABC ventures are, by necessity, precompetitive. In addition, the USABC had some doubts about the potential of lead-acid batteries to meet USABC's performance criteria.

In retrospect, the committee concluded that the exclusion of lead-acid technology from the USABC portfolio had a positive effect on the development of battery technology. In 1992, the manufacturers of lead-acid batteries and their suppliers formed their own consortium, the Advanced Lead Acid Battery Consortium (ALABC), which has an applied research program aimed at making lead-acid technology competitive with the battery technologies being developed under the USABC initiative. The ALABC made significant technical progress between 1992 and 1997 with very little government funding, especially toward achieving the ALABC targets of specific energy, cycle life, recharge time, and purchase cost (Moseley, 1997).

Role of Small Businesses

A number of small businesses have participated in the USABC program. Ovonic Battery Company has developed a Ni/MH (nickel metal hydride) battery, Yardney Technical Products has developed a low-cost nickel electrode, and EIC Laboratories was a subcontractor to W.R. Grace on the development of a lithium-ion-polymer battery. Other small businesses have supplied or manufactured specialty items to USABC developers (Sutula, 1997).

Nevertheless, the committee believes that, despite commendable efforts by USABC management to foster partnership agreements, some features of the procurement process may have put small companies at a disadvantage in bidding for USABC contracts. For example, the absence of specific, published guidelines for cost sharing could have discouraged the participation of small companies that did not have the resources to assume major costs. Even though in practice the amount of cost sharing was negotiable, the time pressures to get the program under way may have discouraged lengthy negotiations. As a result, the participation of some small, less financially robust companies may have been discouraged. In addition, the scoring process for proposals was heavily weighted in favor of organizations with large-scale manufacturing capabilities and demonstrated financial viability. Thus, small, relatively new organizations with promising technologies did not have much chance of participating.

The committee recognizes that the circumstances of the USABC initiative, including the very aggressive time frame for commercialization, might have precluded the participation of small companies that were unable (or unwilling) to

---

[2] Lead-acid batteries are incorporated in EVs being offered commercially by USABC members, such as the Chrysler EPIC minivan, the Ford Ranger light-duty truck, and the General Motors EV1 two-seater sedan (S.C. Davis, 1997).

form partnerships with larger organizations.[3,4] However, the committee believes that participation by more small companies—particularly companies working to develop battery technology relevant to EVs—would have been useful. Historical records show that most "disruptive" technologies were begun by small companies or companies that were not in the mainstream (see, for example, Christensen, 1997). The limited markets for EVs could be attractive to smaller companies, although these markets would not be big enough to meet the commercialization requirements of the 1990 CARB mandate. Also, smaller companies could increase the market penetration, thereby helping to amortize the costs of R&D and the establishment of production facilities.

Commercial Considerations

The committee questioned whether the cost-shared development of midterm batteries would be viewed by many companies as a commercially viable endeavor. Midterm batteries were not expected to capture a sustainable share of the market for EV batteries or even to contribute to the development of long-term systems, and it is not clear when midterm batteries would be displaced by long-term batteries in vehicles. (See chapters 4 and 6 for a discussion of whether midterm battery technologies would be able, after further development, to meet the long-term requirements.) Thus, the opportunities to amortize the development and commercialization costs of a midterm battery may be very limited. If the sales window for EVs with midterm batteries was very short, or was eliminated or delayed (as actually happened), contractors would have little commercial incentive to commit to developing a midterm battery.

Some companies may not have been interested in participating because the Phase I RFPIs were not confidential, despite the fact that the information requested was generally considered proprietary. The RFPI stated that "notwithstanding proposer's markings to the contrary, all information submitted in response to a consortium RFPI shall be treated on a non-confidential basis" (USABC, 1991). Companies with relatively mature battery technologies, which had been developed at considerable expense over many years, were understandably concerned about revealing proprietary information. Negotiations to handle potential problems could have been arranged but would have delayed the project selection process, which was already under severe time pressure.

From the point of view of an industrial supplier, the terms listed in Section VII (Purchase Order Terms and Conditions) of the RFPI would seem restrictive.

---

[3]Ovonic Battery Company and GM have formed a joint venture (GM-Ovonic) to manufacture Ni/MH (nickel metal hydride) batteries.

[4]DOE's Exploratory Technology Research program creates opportunities for small businesses involved in advanced battery research through the Small Business Innovative Research program. The USABC and USABC developers were informed of the results of Phase I and Phase II awards under this small business program.

The seller must commit resources through cost sharing and then turn over all intellectual assets to the buyer. The seller has no guarantee that it will receive a contract to manufacture the system it has developed.[5] (In practice, contracts and agreements were negotiated to provide better incentives for participation. The language in the RFPI should be reconsidered by any follow-on program to the USABC.) The seller's freedom to enter other markets does not fully remove this concern, because a battery that had been optimized for one purpose could probably not be used in another application without extensive and costly redesign.

Science Versus Technology

The committee wishes to make two observations about the development of technology and the development of new science in the context of the USABC initiative. First, the committee acknowledges that the USABC program in Phases I and II was necessarily a technology development program with little room for new science. This was an inevitable consequence of the short CARB timetable for the production of marketable vehicles. Second, the committee strongly believes that new science is essential for the long-term health of the battery industry in the United States and probably is essential for the successful development of practical power systems for electric and hybrid vehicles. The committee does not fault the USABC for not emphasizing new science in its program but believes (1) that DOE should preserve its battery-related scientific research programs in addition to participating in the USABC and (2) that any successor program to the USABC should include relevant new scientific research.

The expectations described in the RFPI clearly indicate that the USABC program is an engineering, rather than a science, initiative, although the proposal evaluation criteria—if applied rigorously—are less definitive. Although the capability of the proposed technology to meet the applicable criteria was an important factor in the decision-making process, the USABC also gave considerable weight to other factors, such as ownership of the proposed technology, high-volume manufacturing process capability, cost sharing, and the financial viability of the proposer's company/team. A large corporation with high-volume manufacturing capabilities and a willingness to share costs with a strong patent could be given a very positive evaluation, even though the patent might cover an impractical or long-range battery technology. In practice, the project selection process was not biased, suggesting that the background knowledge of battery development supplied by DOE and others was a valuable complement to the quantitative proposal evaluation criteria.

---

[5]Section VII.9d of the RFPI states: "At Buyer's request, Seller shall grant to Buyer, its Partners and Partner Associated Companies designated by Buyer, a nonexclusive, paid-up license to make, have made, use, have used and sell under any patents owned or controlled by Seller which cover any application of the technology embodied in the information or data Seller acquires or develops in the course of Seller's activities hereunder" (USABC, 1991).

Because of the very ambitious USABC timetable that required production of a midterm battery in less than three years from the date of the RFPI, the committee assumed that only proposals from companies that had significant technology in hand would receive serious consideration. Battery systems for which there was a scientific base but no supporting technology appeared to be doomed to failure because of the short time frame. With one exception, the committee found that the project selection process adopted by the USABC did focus on battery systems for which technology already existed. Somewhat surprisingly, however, the USABC funded W.R. Grace to develop a lithium-ion battery with a polymer electrolyte, a new system that had not reached the prototype phase of development (see further discussion in Chapter 4).

Thus, on the one hand, the technical challenge was great, and new science was not likely to result in an acceptable product within the time constraints. On the other hand, because the performance and cost requirements were so demanding, the contributions of new science could have been substantial (if the time schedules had been less stringent). If a new battery initiative is being contemplated, the committee believes it should have a better balance between new science and technology development.

FINDINGS

Finding 1. The original objectives of the USABC program led to the establishment of goals that represented great challenges to existing battery technologies. When the targets of pilot-plant production of a midterm battery and demonstration of design feasibility for a long-term battery by 1994 were not met, commercialization goals, which are less demanding than the long-term goals, became the principal focus of Phase II of the program.

Finding 2. The assumption that an EV that fulfills the long-term goals will be fully competitive with an ICE vehicle is not realistic. In range, performance, and recharging characteristics, a battery-run ZEV cannot match an ICE vehicle.

Finding 3. Analyses of the comparative costs of EVs and ICE vehicles—such as the analysis conducted by the USABC—are very sensitive to assumptions about infrastructure and power costs. Thus, achieving the long-term goals may not result in commercially viable EVs that are competitive with ICE vehicles.

Finding 4. Battery systems developed to meet the midterm goals were not expected to gain a sustainable share of the market for EV batteries or even necessarily to contribute to the development of battery systems to meet the long-term goals. The principal purpose of developing midterm batteries was to meet the original CARB requirements for ZEV market share in the late 1990s and establish a market share for EVs. Although this marketing objective has been changed, the development of midterm batteries has been useful in providing power sources for the development of EV technology.

Finding 5. The criteria used by the USABC to evaluate proposals and award contracts were heavily weighted in favor of large organizations with high-volume manufacturing capabilities. The criteria also reflected an aggressive schedule for the commercialization of EV batteries, which left little room for the development of new science.

Finding 6. Small companies with limited resources were at a disadvantage because of the requirements in the RFPI for cost sharing, demonstrated financial viability, and high-volume manufacturing capabilities. The USABC's encouragement of partnerships among interested parties did help to counteract this disadvantage, although promising innovative technologies under development by small companies may still not have been given adequate consideration. As a result, small companies, including those working to develop battery technology relevant to EVs, have not been active participants in the USABC program, although DOE has created opportunities for small businesses to pursue R&D on batteries through the Small Business Innovative Research program.

Finding 7. Decisions to invite or exclude Asian participation in the USABC program have been inconsistent, but the committee could not determine whether there has been—or is likely to be—any resulting impact on battery development.

Finding 8. The decision by the USABC to exclude lead-acid battery technology from its program had a positive impact on the development of this technology. The Advanced Lead Acid Battery Consortium undertook its own applied research program, which has resulted in significant improvements in lead-acid battery technology.

4

Technical Progress

This chapter provides a review of the technical progress made in USABC battery technology R&D programs through the end of 1997. The committee's review addresses work done in Phase I, commitments for Phase II, and work in progress through 1997. The chapter is organized by battery technologies, namely, Ni/MH (nickel metal hydride), lithium, and fused-salt battery systems. The committee assessed the potential of each technology for EVs on the basis of progress made by the USABC and the technical goals and objectives discussed in Chapter 3. The committee also identified the limitations of each technology.

NICKEL METAL HYDRIDE BATTERIES

The Ni/MH battery uses a nickel positive electrode (developed a hundred years ago) and a hydrogen-absorbing alloy negative electrode (developed in the 1970s). The packaging and most of the manufacturing processes are similar to the nickel-cadmium (Ni/Cd) battery. Metal hydrides used for the negative electrode include AB_5-type alloys based on rare earths and nickel, which are being developed by SAFT, and AB_2-type alloys based on nickel and vanadium, which are being developed by GM-Ovonic. The metal hydride is an environmentally friendly replacement for the toxic cadmium in the Ni/Cd battery; the Ni/MH battery also has a higher specific energy than the Ni/Cd battery. Small Ni/MH batteries are used extensively in portable consumer electronics, such as cellular phones, laptop computers, and VCRs. Technical information on the three USABC Ni/MH battery programs is summarized in Table 4-1.

TABLE 4-1 Technical Summary of USABC Nickel Metal Hydride Battery Projects (as of November 1997)

Principal Developer(s)	Technology	Status	Current Activities
General Motors-Ovonic (GMO)	Midterm battery with negative electrode based on AB_2-type alloys	Battery packs meet midterm criteria for power density and volumetric energy density but not for specific energy. Cycle life for battery packs is close to midterm requirements; calendar life has not been determined. Cost reduction is a major challenge.	Electrode materials are being developed to reduce cost and increase specific energy, including high-risk work on MgNi-based alloys as alternatives to AB_2-type alloys for negative electrode.
SAFT America, Inc.	Midterm battery with negative electrode based on AB_5-type alloys	Battery packs meet midterm criteria for power density and volumetric energy density but not for specific energy. Cycle and calendar life have yet to be determined. Cost reduction is a major challenge.	Work is focused on reducing cost through optimization of materials and using fewer high-cost additives.
Yardney Technical Products	Low-cost, fiber-based, pasted nickel electrode	Performance of electrode in GMO and SAFT cells is disappointing compared to performance of electrodes developed in house.	Project has been terminated.

GM-Ovonic

In May 1992, the USABC awarded Ovonic Battery Company a contract to develop a Ni/MH battery for EVs based on Ovonic's proprietary metal hydride AB_2-type alloy. The goal was to meet the CARB mandate for the commercial introduction of EVs (with subsidies) in 1998, i.e., the system was to be developed as a midterm battery. In 1994, Ovonic Battery Company and GM entered into a joint venture, GMO (GM-Ovonic), to manufacture the battery developed under the USABC contract. The original contract called for a three-year program consisting of three phases[1] with increasing cell performance and module and battery-pack deliverables. The deliverables were to undergo testing at GMO, at Argonne National Laboratory (ANL) under a CRADA, and in EVs developed separately by Chrysler, Ford, and GM.

Prior to the USABC contract, Ovonic had developed 25-Ah to 35-Ah prismatic cells that could deliver more than 600 cycles and had a specific energy of 54 Wh/kg. The purpose of the first phase of the USABC project was to duplicate the performance of these cells in multicell strings. The second and third phases focused on developing larger cells with higher specific energy. Deliverables during this program were multicell modules and battery packs, with 4 to 10 modules, for both laboratory-scale testing and vehicle testing. The modules and battery packs were expected to meet the midterm goals of the USABC. GMO has produced individual cells that meet the midterm goals, but their specific energy is lower at the module and battery-pack levels.

The modules and battery packs are being tested at ANL and in EVs by the three major U.S. car companies. The close interaction between the car companies and GMO in the early stages of battery development has led to improvements in the design (e.g., pressure release vents and terminal seals) and packaging of the cells and modules and to a better understanding of systems-level integration. The best packs deliver 450 to 600 cycles, which is close to the midterm requirements. The battery packs lose their power density with cycling and, therefore, do not meet the USABC calendar life requirements. The early EV tests raised questions about thermal control of the battery, so GMO made some preliminary tests of cell performance at different temperatures. The results of these tests showed that the modules will require active cooling, which will affect the specific energy of the battery. Preliminary failure modes and effects analysis (FMEA), safety, and charge control issues have also been addressed.

In Phase I of the USABC program, GMO has increased the specific energy of Ni/MH cells from 50 Wh/kg to 70 Wh/kg. The increase in specific energy and

[1]The three phases of the Ni/MH program at GMO were all part of the Phase I USABC program. The interim phases were created for organizational purposes and had increasingly demanding goals. In June 1994, a fourth phase was added focused on cost reduction and establishing a pilot manufacturing facility.

performance of the battery resulted from significant changes in the electrode materials, including changes in the chemical composition and manufacturing processes for the positive and negative electrodes. These changes were the result of materials research conducted parallel to the development of larger cells. Materials changes in the cell designs initially caused premature failures and a loss in power density after cycling. After several iterations, however, GMO was able to rectify most of these problems.

Cost analyses were conducted throughout the program. The final cost estimate for GMO's Ni/MH battery for production of 20,000 packs per year is more than 1.5 times the midterm goal of \$150/kWh. These values are close to the estimates by SAFT for its Ni/MH battery and, in both cases, a large part of the cost (60 percent) is for materials. High cost is obviously a major problem for the acceptance of Ni/MH technology in EVs. The cost of a product is very difficult to estimate when significant materials and design changes are still being made, and the committee has reservations about the validity of the cost analysis for Ni/MH presented at the first committee meeting (Rauhe, 1997). The \$150/kWh target is the selling price for the battery, whereas the high cost estimate is for the production cost of the battery. For example, the selling price of small Ni/MH batteries for portable applications is currently higher than \$600/kWh.

In June 1994, a 16-month Phase IV was added to the Ovonic Battery Company contract. The program was amended to include significant work on electrode materials to reduce costs and to increase the specific energy of the battery beyond the midterm goals. This was high-risk research, which was very different from the evolutionary improvements made in Phases I, II, and III of the Ovonic program. However, Phase IV did not result in significant reductions in cost.

SAFT

In December 1992, the USABC entered into a contract with SAFT America, Inc., of Cockeysville, Maryland, to develop Ni/MH batteries based on an AB_5 alloy for a midterm EV battery.[2] The original contract lasted through March 1996. In October 1994, additional funding accelerated parts of the program (i.e., delivery of the first battery pack by the end of 1994) and increased the number of deliverable 40 kWh battery packs to five.

Since the 1970s, SAFT has built Ni/Cd cylindrical and prismatic cells for portable applications, such as laptop computers, camcorders, and power tools. SAFT used a hydrogen-absorbing AB_5-type alloy in the negative electrode of NiMH batteries throughout Phase I of the USABC program. At SAFT, the Phase I USABC program addressed the design of cells and battery packs, cost, FMEA, safety analysis, the development of diagnostic criteria, the development of cells

[2]SAFT America, Inc., is the American arm of the French company, SAFT, which is a well known manufacturer of several kinds of batteries.

with higher specific energy, and the development of high energy density modules. Deliverables included cells and modules with six 40-kWh battery packs to be delivered at the end of the program.

SAFT has made improvements in both the positive Ni electrode and the negative hydride electrode and has produced cells with a specific energy of 70 Wh/kg. Five cells are connected in series (6 V modules) to produce a maintenance-free monoblock with active water cooling and a specific energy of 65 Wh/kg. As many as 56 modules are required to form a battery pack. The modules meet all of the midterm performance criteria except specific energy. Data on improved AB_5 alloys suggest that energy density per module can be increased to approximately 70 Wh/kg. Independent tests have been initiated by Ford, Chrysler, GM, DOE, the Idaho National Engineering and Environmental Laboratory, and EPRI on the battery packs delivered to the USABC, but the cycle life and calendar life of the SAFT battery have not been determined.

The cost of the SAFT and GMO Ni/MH batteries is estimated to be at least 65 percent higher than the midterm goal. The estimate is for 25,000 battery packs per year. SAFT believes that further improvements will bring the cost closer to the midterm goal of $150/kWh. SAFT has presented a detailed plan for the pilot-plant production of battery packs.

Yardney Positive Electrode

In 1994, the USABC awarded Yardney Technical Products a contract to develop a low-cost, fiber-based, pasted Ni electrode for EVs. The goals of the program were to develop a high-capacity electrode at a cost of $43.60/kWh with no more than a 20 percent loss of capacity over 1,000 dynamic stress test (DST) cycles.[3] At the end of the project, Yardney claimed to have produced the high-capacity electrodes with an estimated cost of $56/kWh at production rates of 20,000 battery packs per year and a 7.5 percent loss of capacity after 600 cycles. The electrodes were provided to ANL and the two Ni/MH battery developers, GMO and SAFT, for testing. Both battery companies found that the Yardney electrodes delivered significantly lower capacity than the positive electrodes that had been developed in house. As a result, the USABC terminated the Yardney project.

LITHIUM BATTERIES

Lithium metal, because of its low equivalent weight[4] and its electrode potential, is an attractive candidate material for the negative electrode in high energy

---

[3]A dynamic stress test simulates the pulsed-power charge and discharge environment of EVs and is based on the Federal Urban Driving Test regime. This test provides more accurate battery performance and life estimates than constant-current tests because it more closely approximates the application requirement.

[4]Equivalent weight is the molecular weight per electron delivered. The equivalent weight for lithium is the atomic weight of lithium divided by one.

density batteries. The positive electrode consists of intercalation compounds,[5] such as sulfides and oxides of transition elements, that can reversibly intercalate lithium ions. During the initial development of lithium rechargeable batteries in the 1960s and 1970s, organic solvents containing lithium salts were used as electrolytes. However, the cells had serious safety problems because of the high chemical reactivity of lithium metal in liquid electrolytes. Two solutions to this problem are being investigated by the USABC. The 3M/Hydro-Québec lithium-polymer battery replaces the liquid electrolyte with a solid polymer that contains a lithium salt, which significantly reduces the tendency for reaction between the lithium and the electrolyte. The second approach, in which the lithium metal is replaced by an intercalation compound that has a sufficiently negative electrode potential, such as graphite, is used in the Duracell/VARTA lithium-ion battery. In this battery, no lithium metal is present, and the lithium ion shuttles between the two intercalation compounds. Lithium-ion batteries, also called "rocking chair" batteries, are used extensively in high-end portable consumer products. The W.R. Grace lithium-ion-polymer battery attempted to combine the benefits of the lithium-ion and polymer approaches. Technical information on the USABC's three lithium battery programs is summarized in Table 4-2.

W.R. Grace Lithium-Ion-Polymer Battery

A team[6] led by W.R. Grace was funded beginning in April 1993 for the development of an organic electrolyte-polymer lithium-ion system with a carbon anode and a manganese oxide cathode. Both bipolar and monopolar configurations were developed; however, the bipolar design was abandoned in favor of the monopolar design midway through the project. The system required the development of a new battery type, including the development of new materials, before it could reach the prototype phase. The goal of the project was to provide a proof-of-concept system, including the production of a 20 kWh battery. Based on the

[5]Intercalation or insertion compounds allow the insertion of a guest species, such as lithium, into the host lattice during discharge and subsequently extract the lithium during discharge without changing the structure of the host lattice.

[6]The final team included W.R. Grace, Johnson Controls, SRI International, EIC Laboratories, and USCAR Carbon, Inc. W.R. Grace provided expertise in polymer processing and film manufacturing, as well as some knowledge of lithium-ion battery chemistry. Johnson Controls had manufacturing know-how, as well as expertise in modeling and testing batteries and experience in designing and assembling both monopolar and bipolar designs. EIC Laboratories had a proprietary position in polymer electrolyte technology and experience building lithium battery prototypes. SRI International contributed expertise in novel solid polymer electrolytes; USCAR Carbon, Inc., provided expertise in carbon anode materials. Thus, the final team had a combination of expertise in basic research and proven experience in manufacturing. Work on carbon, cathode materials, and lithium-polymer battery recycling was conducted at Sandia National Laboratories in support of the W.R. Grace and other lithium-polymer programs.

TABLE 4-2 Technical Summary of USABC Lithium Battery Projects (as of November 1997)

Principal Developer(s)	Technology	Status	Current Activities
W.R. Grace	Midterm lithium-ion battery with polymer electrolyte for ambient temperature operation	Significant progress has been made in developing this battery, which had not previously been prototyped, but midterm goals were not met.	Not applicable.
Duracell/VARTA	Midterm lithium-ion battery with organic electrolyte	Midterm performance criteria have been met, although life and cost goals have not been demonstrated. This technology is also of interest for high-power batteries for hybrid vehicles.	Work is being done on the assembly of EV-sized modules that incorporate additional safety concepts; testing and optimization are being done.
3M/Hydro-Québec	Long-term lithium-polymer battery	Midterm performance criteria have been met, although life and cost goals have not been demonstrated. This technology comes closest to meeting long-term goals. A reliable thermal management system is needed because the battery does not function below 40°C.	The company hopes to demonstrate the viability of meeting interim commercialization performance goals at the 1.7 kWh modular battery level. A scale-up process is being developed for manufacturing a cathode/solid polymer electrolyte laminate.

state of the technology, this goal was modified to provide only a demonstration of the viability of the technology in a full-scale building-block module.

Models were developed to facilitate cell and battery design and to predict thermal behavior. Safety and reliability issues were addressed through thermal modeling and the development of a preliminary system FMEA. A study of production costs was also partially completed. By the end of the program in 1996, the three-year program had yielded laboratory prototype cells and a pilot-plant facility, which had produced more than 4,500 meters of electrode material. Scale-up of the material width to a size required to build cells and modules for EV application was progressing.

Although significant progress was made, the goals were not met. Cost, cycle life, energy and power density, and electrical controls were still outstanding issues. Significant development of the materials technology would have been necessary before the long-term goals could be met. For business reasons, W.R. Grace decided not to pursue further development of this technology. The USABC has indicated that the decision not to pursue Phase II was a joint decision by the contractor and the USABC.

Duracell/VARTA Lithium-Ion Battery with Organic Electrolyte

The Duracell/VARTA lithium-ion battery proposal was solicited and funded in January 1995, near the end of Phase I. The objective was to develop a midterm battery, although, in the opinion of the USABC, the technology could eventually approach the long-term goals. Unlike the 3M projects, this lithium-ion technology has a liquid organic electrolyte. (The USABC did not initially pursue a liquid organic electrolyte system, but in 1995 liquid electrolyte systems became attractive for use in hybrid vehicles.) The Duracell/VARTA battery is of interest to the Partnership for a New Generation of Vehicles (PNGV) program as well as the ZEV targeted in the USABC program.

The comprehensive Duracell/VARTA proposal reflected the long-term experience of both Duracell and VARTA in designing and marketing battery products for a variety of challenging applications. Planned activities included the development of electrical monitoring and control, thermal management, and safety testing and simulation, as well as FMEA. One of the deliverables was a proposal for a manufacturing process that incorporated statistical process controls. Cost analysis, disposal, and recycling were also addressed in the plan. Large prismatic cells with cobalt oxide cathodes had been developed independently by Duracell/VARTA prior to their USABC proposal.

Under the USABC contract, a lithium-ion battery with a manganese oxide cathode was scheduled to be developed. The initial program was redirected to incorporate further abuse testing, including a possible change in cell design and modified materials. The first phase of the program was completed in January 1997.

At the end of 1996, a Phase II contract was established that extended the program for 24 months. Phase II focuses on the incorporation of nine previously identified concepts to make EV-sized modules safe and includes hardware assembly, testing, and optimization. Lowering costs by optimizing materials processing and cell assembly was also planned. Testing to date suggests that the midterm performance criteria will be met, although cost and life goals have yet to be demonstrated. The Duracell/VARTA program has conducted effective module and systems engineering tests for safety and reliability.

3M/Hydro-Québec Lithium-Polymer Battery

The 3M/Hydro-Québec project is the only one presently being funded to meet the USABC long-term performance criteria. At the time the contract was signed in December 1993, the basic technology was already reasonably mature. Hydro-Québec had been working on its development for approximately 15 years, at a cost of $40 million (Canadian), for energy storage applications. The battery has a lithium anode, a solid polymer electrolyte, and a vanadium oxide cathode and operates over a temperature range of 60 to 100°C.

The USABC considered the lithium polymer technology to be the best candidate for meeting the long-term goals by 2003. The USABC provided a significant boost to its development, not only by providing additional funding and a possible market, but also by helping to fill in several gaps in the development program. Teaming up 3M (which had submitted a separate proposal), ANL, and Hydro-Québec in a three-way CRADA provided a number of benefits. The participation of 3M brought world-class R&D and manufacturing know-how in thin-film polymers to the partnership, as well as addressing the USABC goal of developing a U.S. battery manufacturing base. ANL was brought into the project to provide expertise in battery design, modeling, and testing, as well as materials development for advanced positive electrodes.[7]

The USABC also brought battery technologists together with EV experts to provide expertise in systems integration and testing. Although this expertise was important to all of the USABC projects, it was crucial to 3M and Hydro-Québec, which, unlike Johnson Controls and VARTA, had no previous experience designing batteries for the automotive industry. The committee noted that the 3M project leader was very positive about working with the automotive companies through the USABC.

The goals of the two-year Phase I segment of the 3M/Hydro-Québec project were to meet or exceed the USABC's midterm performance criteria; to demonstrate design feasibility for lithium polymer battery technology; to explore the properties and limitations of advanced materials used in the particular system

---

[7]The lithium vanadium oxide used in the 3M/Hydro-Québec battery is a joint invention of 3M and ANL.

under development; and to evaluate configurations that could meet USABC's long-term goals. In the committee's view, the project was well planned, which was apparent in the technology transfer from Hydro-Québec to 3M early in the project. Manufacturing facilities were built, and the development of manufacturing processes was begun immediately at both 3M and Argo-Tech, a subsidiary of Hydro-Québec. Tasks were defined to develop models for battery performance and cost and to address safety and reliability, electrical control, thermal management, and systems integration issues. The management practices and statistical design of experiments were also good. The project leader at 3M has extensive experience in managing large government programs, which, the committee believes, contributed substantially to the success of the project.

By the end of Phase I, a 100-Wh lithium polymer battery composed of a parallel arrangement of five 20-Wh prismatic cells was delivered. Projected costs for the battery were significantly higher than the interim commercialization target. In addition, the battery cell balance and control strategy had been developed, and the preliminary battery design had been completed. Significant safety testing had also been completed at the 100-Wh cell level. The performance of the battery was projected to exceed the midterm goals. The most significant issues that still needed to be addressed to meet the interim commercialization goals were cycle life, rate of discharge, and cost. However, the battery could not meet power requirements at temperatures below 40°C because the solid polymer electrolyte has low ionic conductivity. A reliable thermal management system must be developed for vehicle start-up after the battery has cooled down.

The objective of Phase IIA is to demonstrate the viability of the lithium polymer battery technology in meeting the USABC interim commercialization performance goals at the 1.7-kWh modular battery level. In June 1997, the USABC awarded additional funding for reducing cost and improving performance. This Phase IIB, which will last for two years, will culminate in the first production series of full-sized battery packs.

At the beginning of Phase IIA, electrode and electrolyte materials were still being developed and selected to establish stable performance. Two 12-V, 760-Wh submodules were being delivered. Scale-up, delivery, and safety testing of the technology with 1.7-kWh modules and minipacks were scheduled for June 1997; another six minipacks were scheduled to be delivered in December 1997. The process for making the cathode/solid polymer electrolyte laminate is now being scaled up. At the end of 1998, if Phase IIB has been successful and the original equipment manufacturers of EVs make a commitment to purchase a given number of batteries, 3M and Hydro-Québec plan to expand their manufacturing capability at their own expense to support a market launch of the lithium-polymer technology. A final design freeze is scheduled for July 1999. A total of nearly $85 million will have been invested by the USABC, 3M, and Hydro-Québec to bring the lithium polymer battery concept to the prototype level (Letourneau et al., 1997).

Based on the most recent quarterly progress report (April through June 1997), the committee determined that the 3M/Hydro-Québec project appears to be on target to meet its deliverables. The program has made significant progress in the development of manufacturing capabilities and has assembled and tested an impressive number of prototypes. The program appears to be well managed, and the contractors appear to be committed to the program.

FUSED-SALT BATTERIES

Two fused-salt battery systems were chosen for development by the USABC: sodium-sulfur (Na/S) and lithium-iron disulfide (Li/FeS_2). Both are elevated temperature systems. The Na/S system operates at a nominal temperature of 325°C, and the Li/FeS_2 system operates in the 450 to 500°C range, depending on the composition of the fused salt. The Na/S system uses two ionic conductors in series: a ceramic solid electrolyte system and a sodium polysulfide salt, which is a liquid at operating temperatures. The ceramic is a sodium-ion conductor that mechanically separates the positive and negative reactants and also limits the conductivity to sodium ions only, thus preventing cell self-discharge.

Silent Power Sodium-Sulfur Battery

In August 1993, a contract was signed with the American office of the Silent Power group, which was owned at that time by the German RWE group. RWE had purchased the background experience of the British company, Chloride Silent Power, Ltd., which had been working on the Na/S system for more than 20 years. The USABC contract was structured to exclude all cell development and to focus on generic battery designs, features, and processes that could be applied to any candidate Na/S technology. The Na/S technology was also being investigated by other groups in Germany, Japan, and China.

The Silent Power project addressed the areas of thermal management, mechanical assembly (especially reliability and cost), systems integration, and controls. Various approaches to cooling were studied, including cooling by air or liquid (an oil) and by varying hydrogen pressure, referred to as variable conductance. Variable conductance was investigated at the National Renewable Energy Laboratory. Liquid cooling was studied, with the variable conductance as a backup.

A major goal of the American office of Silent Power was to provide an improved design of the Na/S battery that could be used for EVs, and the planar module design did meet the USABC requirements and passed several tests. An 840-cell battery was subjected to extreme failure testing, including dropping the battery onto a steel post, which severely distorted the battery enclosure and caused the noncatastrophic failure of 21 cells. The battery was also subjected to prolonged vibration. Cost analyses showed that the midterm USABC price goals could be met, given adequate levels of production.

The automotive Na/S battery designed for the USABC involved a new cell design that was being developed by the parent company, Silent Power RWE. At the end of the USABC contract, the cells had not yet demonstrated required life expectations. For this reason, the USABC decided to terminate the project. Under the USABC contract, the cell had been redesigned and tested by Silent Power in England and Germany; those results were made available to the USABC, although Silent Power RWE retained the rights to the technology.

This USABC project was difficult to assess because most of the requisite data were supplied by Silent Power RWE. The committee was not provided with the program design or the results of prolonged testing of the new cell design. The American office of Silent Power appears to have directed its efforts toward crucial design areas and to have made significant progress toward meeting its objectives.

Lithium-Iron Disulfide Battery Project

A contract between the USABC and SAFT America, Inc., for the development of an advanced, high-temperature Li/FeS_2[8] battery for EV propulsion was signed in December 1992. A CRADA was established between SAFT and ANL, which has a long history of working with the Li/FeS_2 system. SAFT took the lead in module development, and ANL assumed responsibility for technical support.

Technical issues for the Li/FeS_2 system include the loss of capacity caused by the migration of reactants through the separator, the composition of the separator, the optimum composition of the fused salt, and the construction of long-lived corrosion-resistant seals. Solutions to these technical problems are likely to increase the cost of the battery.

Studies of both the positive iron disulfide (FeS_2) electrode and the negative lithium alloy electrode were conducted, and the performance of both was improved. The paste electrolytic separator, based on magnesium oxide, was also studied and improved. The fused-salt electrolyte itself was investigated for various compositions, and an optimum ternary composition involving lithium ion as the only cation was found. This composition lowered the melting and operating temperature and increased ionic conductivity.

A good deal of effort was directed toward improving seals and cases. At the elevated operating temperatures of the battery, the salt components of the cell tend to be aggressive corrosion agents, which drastically shortened the lifetimes of the seals and cells. Studies were done on the packaging design of cells and

---

[8]At the operating temperature of the cell, lithium metal (which has a melting point of 186°C) is a liquid and is difficult or impossible to confine to the region of the cell intended for the negative electrode. Using a lithium alloy that is a solid at operating temperatures is a convenient way to confine the lithium. Thus, this battery is actually a lithium alloy/iron disulfide system.

modules and the temperature management of the resulting package. Studies were also done on managing the consequences of overcharging and overdischarging.

At the conclusion of the contract, the cells were capable of 200 Wh/kg and nearly 500 W/kg. The project was mutually terminated because of shortfalls in cycle life, calendar life, and cost.

FINDINGS

General Finding

Finding. The USABC brought together battery technologists and experts in EVs, thereby providing a valuable systems integration and testing focus to the battery development programs.

Nickel/Metal Hydride Batteries

Finding 1. Significant improvements have been made in Ni/MH battery technology in the programs coordinated by the USABC, particularly in developing large cells and batteries for EVs. Furthermore, USABC projects have led to a better understanding of design, manufacturing processes, and cost issues for Ni/MH systems and have established processes for the pilot production of electrodes, cells, and modules. However, significant work on battery systems-level integration and evaluation remains to be done before the technology will be roadworthy.

Finding 2. Ni/MH technology meets the midterm targets for power density and volumetric energy density but does not meet the most critical midterm criteria for specific energy and cost. The calendar life of Ni/MH batteries has yet to be determined.

Finding 3. Because the cost of the Ni/MH batteries is very high, the USABC has decided to focus on reducing the cost in Phase II. It remains to be determined whether the costs can be reduced enough to meet the USABC's interim commercialization goals.

Finding 4. The USABC's decision to develop two competing Ni/MH technologies based on AB_2- and AB_5-type hydrogen absorbing alloys (GMO and SAFT technologies, respectively) provided a good insurance policy in the event that either of the technologies failed to meet program objectives.

Finding 5. The Yardney program to develop a low-cost nickel electrode suffered by not interacting with battery manufacturers early in the program. Because electrode and cell designs could not be satisfactorily integrated, the performance of the battery was disappointing.

Lithium Batteries

Finding 1. Substantial progress has been made on lithium batteries, although all systems failed to meet their cost objectives. The 3M/Hydro-Québec lithium polymer battery and the Duracell/VARTA battery both met the midterm performance requirements, although cycle and calendar life have not been demonstrated yet because of the long test times. Preliminary results suggest that they may not meet the long-term goals.

Finding 2. The Duracell/VARTA battery may be able to meet the midterm goals, although not in the midterm time frame, and may offer opportunities for further development to meet USABC interim commercialization goals; there is a separate program for meeting PNGV requirements for hybrid vehicles. Some safety issues remain to be addressed.

Finding 3. The 3M/Hydro-Québec battery is projected to come close to meeting the interim commercialization goals and comes closest of all the technologies to meeting the long-term goals. Further work will have to be done on the recycling or disposal of spent batteries, safety testing, and performance testing under extreme temperatures and stressful-use conditions. A reliable thermal management system will have to be developed for vehicle start-up after the battery has cooled down.

Finding 4. Final design freezes for the midterm Duracell/VARTA lithium-ion battery and the long-term 3M/Hydro-Québec lithium polymer battery are planned for mid-1998.

Finding 5. The W.R. Grace team made progress in developing its lithium-ion polymer technology from an embryonic stage, as the USABC requested. Nevertheless, the technology was not ready for full-scale development within the USABC time frame.

Finding 6. The 3M/Hydro-Québec lithium-polymer battery program has made substantial technical progress in the context of the USABC goals and time frame.

Fused-Salt Batteries

Finding. The SAFT-ANL project to develop a Li/FeS_2 battery was well organized, but not enough emphasis was placed on known problem areas for high-temperature fused-salt systems, namely, cell cycle life and calendar life.

5

Management Model and Oversight

Beginning in 1980, Congress enacted a series of laws to renew, expand, and strengthen cooperation between the federal laboratories and private industry. The Stevenson-Wydler Technology Act of 1980 (P.L. 96-480) made the transfer of technology developed under federal programs a national priority. This act was amended by the Federal Technology Transfer Act of 1986 (P.L. 99-502), which authorized government-operated laboratories to enter into cooperative research and development agreements (CRADAs) with nonfederal parties to conduct specific R&D. In recent years, a number of government-industry cooperative R&D programs have been established, including PNGV (Partnership for a New Generation of Vehicles), Sematech, and USABC, to name but a few.

Most federal agencies have used CRADAs as a means of conducting R&D projects of mutual interest with industry (see, for example, Wells, 1993; GAO, 1994; Branscomb and Keller, 1998), although no single management structure has emerged as the model most likely to succeed. To the surprise of the committee, a search of the literature did not reveal any attempts to integrate lessons learned from the management of diverse government-industry R&D programs. A comprehensive review of these programs was beyond the scope of the present study, but at the request of the study sponsor, the committee attempted to identify some lessons from the USABC that might benefit government-industry R&D partnerships in the future.

This chapter provides the committee's assessment of the USABC's monitoring and evaluation of contractor performance, followed by a review of lessons learned during the USABC program and a discussion of some common elements between the USABC, PNGV, and Sematech.

EFFECTIVENESS OF PROGRAM MANAGEMENT

Written material provided to the committee, together with presentations made at committee meetings (see Appendix B), indicate that the three major U.S. automakers, Chrysler, Ford, and GM, have directed the USABC program, and the management structure (see Chapter 2) vests authority for critical decision making in their representatives. DOE contributes a large part of the funding for each project (50 percent in Phase I, 45 percent in Phase II), and DOE's decision-making role is carried out through a combination of consensus management and oversight functions (see Chapter 2). DOE is not represented on the partners' committee, and DOE representatives at management committee meetings are nonvoting participants, but they do provide oversight. DOE representatives are full participants in the TAC and working groups. DOE's formal oversight of the USABC program is spelled out in the cooperative agreement between DOE and the USABC.

EPRI also plays a subordinate role to the automakers. EPRI is not represented on the partners' committee, and representatives of EPRI were nonvoting participants on the management committee in Phase I. In Phase II, however, EPRI has a voting member on the management committee. In addition, an EPRI representative from the TAC is responsible for managing one of the major battery development programs (USABC, 1996). The USABC secretary and business manager is also from EPRI.

In practice, the relationships among representatives of the automakers, DOE, and EPRI appeared to the committee to be largely collegial rather than directive. For example, although all of the voting members of the management committee are representatives of the automotive companies, any course of action opposed by DOE was not likely to be adopted. In other words, the management process is consensual.

The committee strongly supports the involvement and leadership role of managers from the three U.S. automakers in the USABC. Their participation provides the real-world experience, knowledge, and judgment required for this complex undertaking. A large-scale transition from the ICE automobile to EVs—as anticipated by the 1990 CARB mandate—will certainly require an understanding of the demands of the American driving public, including expectations of performance and safety. Representatives of the automakers are well positioned to provide this understanding, and the USABC goals were derived from market-driven vehicle specifications set by Chrysler, Ford, and GM.

The committee was favorably impressed by the USABC's use of many proven industry practices for managing projects, controlling costs, auditing, and protecting intellectual property. The USABC has established effective mechanisms to ensure that the original statements of work for USABC programs are being followed or are adjusted according to established procedures. As a result, R&D has been focused on critical technical issues. The industrial leadership model has had a very positive influence on the management of the consortium. Decision making has been effective and timely, and projects that have not met defined milestones have been terminated.

In the USABC management structure, program managers pay a critical role. Established industry practices require that program managers be experienced personnel capable of evaluating adherence to schedules and assessing progress toward meeting cost and performance goals. Information provided to the committee indicated that USABC program managers are appropriately qualified. However, because their role in the consortium's activities is critical, the committee believes that the selection of program managers should be subject to external peer review.

The program managers review the technical status of programs at TAC meetings on a quarterly basis. (Each work group consists of the managers of all programs.) The committee was informed that during Phase I there were 10 to 15 program managers, corresponding to the number of ongoing development programs. Smaller work groups managed individual programs. The committee was concerned that program reviews by the current work group may not be effective because of the collegial relationship among the program managers. Although their reviews may be satisfactory, the committee believes that the consortium as a whole would benefit from external reviews.

The committee's principal concern regarding the effectiveness of the USABC management structure was in the area of peer review. Both USABC peer reviews and critical reviews are performed by members of the TAC and are the basis for communicating status and issues to the management committee. Thus, USABC peer reviews are done by personnel within USABC's management structure. Independent, outside experts do not participate—in contrast to the peer review process of other government-supported R&D programs. The committee is concerned that the USABC peer review process may not be sufficiently objective because program managers are likely to defend developers for whom they have oversight responsibility.

LESSONS LEARNED

A review of lessons learned from the entire process associated with the formation of the USABC was conducted in 1993 (Abacus Technology Corporation, 1993). The purpose of this review, which involved participants from industry and DOE, including the national laboratories, was for both government and industry to obtain a better understanding of the issues and improve the consortium model for future government-industry cooperation. The USABC's negotiations of agreements with subcontractors for the development of advanced batteries, as well as several CRADAs, were evaluated. Seventeen issues were identified, covering areas such as the approach to intellectual property in partnerships, tailoring negotiation processes to accomplish DOE policy objectives, and the

CRADA review and approval process. The major findings of a subsequent review of DOE's implementation of the lessons learned (Abacus Technology Corporation, 1996) are listed below:

- Intellectual property agreements, which have been shortened and simplified, are significantly easier to implement. DOE fashioned its new patent waiver regulations after the Federal Acquisition Regulations (FARs), which are standard throughout the federal government.
- The process of negotiating USABC agreements has been significantly improved by increasing the quality and level of communication between DOE and the USABC and by the USABC designating a full-time business manager in 1994.
- The process of reviewing and approving contracts and agreements has been significantly improved by lifting the requirement for approval of subcontract terms and conditions, while retaining the requirement that cost and pricing information be submitted to the contracting officer for preliminary review and approval.
- The revised intellectual property agreements have also streamlined the CRADA review and approval process; the technical review and approval of subcontracts remains a responsibility of DOE headquarters.
- The agreements were streamlined by retaining the applicable cost principles (FAR Part 31) and eliminating the application of cost accounting standards in cooperative agreements.

A further assessment of the USABC conducted by the U.S. General Accounting Office (GAO) determined the USABC's progress toward reaching its long-term and midterm goals; the funding that had been spent as of FY 1995 and the additional funds, if any, that will be needed; and the role of DOE in managing the consortium (GAO, 1995). The GAO found that DOE had not adequately responded to the "lessons learned" report (Abacus Technology Corporation, 1993) and had, therefore, missed some opportunities to make the consortium, and similar cooperative efforts in the future, more efficient.

Following discussions with representatives of the USABC and DOE, the committee concluded that DOE has now addressed the significant issues raised by the "lessons learned" report and has acted upon the ones that are within its jurisdiction. For example, DOE has adopted new procedures that have resulted in the execution of a new cooperative agreement with the Advanced Reactor Corporation and a second cooperative agreement with the USABC for high-power energy storage technology in support of PNGV. Both agreements were executed in significantly less time than the original agreement for the USABC (Heitner, 1997). In the committee's judgment, DOE would benefit by taking advantage of this experience when launching new collaborative government-industry R&D programs, including a possible follow-on program to the USABC.

Based on the collective knowledge of committee members of DOE programs, the committee had the impression that several different practices are adopted across the department when setting up cooperative agreements and that little effort has been made to standardize these practices or make them user friendly. In the committee's view, the issue of contracting reforms could be addressed under the auspices of the DOE R&D Council or the Laboratory Operations Board.

Rapid implementation of the cooperative agreement between the USABC and DOE was hampered by major differences between government and industry practices with respect to the development and commercialization of technology. Although none of these differences was a "showstopper," their resolution required management attention that detracted from the efficient and expeditious implementation of the technical development program and made an already challenging schedule even more difficult to maintain. For example, the original equipment manufacturers in the automotive industry typically request that suppliers fund or share the cost of developing automotive component technology but do not require access to the technology. In contrast, DOE requires access to both pre-existing and newly developed technology (Abacus Technology Corporation, 1993). The first two years of the USABC's existence were largely taken up with the logistics of establishing the consortium and its operating principles and resolving issues relating to intellectual property. Nevertheless, the committee did not observe any adverse effect of this long induction period on the eventual selection of midterm and long-term battery technologies.

OTHER MANAGEMENT MODELS

The committee considered certain aspects of two other government-industry collaborative R&D programs—PNGV and Sematech—to determine if DOE has benefited, or could benefit, from their management experience. The committee recognizes, however, that PNGV was formed in September 1993 and thus had the benefit of the experience of the USABC's initial experience.

Partnership for a New Generation of Vehicles

The PNGV program is a cooperative R&D program between the federal government and the USCAR, which is made up of Chrysler, Ford, and GM. The decade-long program was instituted on September 20, 1993, by President Clinton with the goals of (1) significantly improving national competitiveness in manufacturing; (2) implementing commercially viable innovations from ongoing research on conventional vehicles; and (3) developing a vehicle to achieve up to three times the fuel efficiency of today's comparable midsized vehicles (e.g., the Concorde, Taurus, and Lumina) while maintaining or improving current performance levels, size, utility, and total cost of ownership and meeting or exceeding safety and emissions requirements.

Government funding for PNGV is used primarily for development of high-risk technologies (Goal 3). The three USCAR partners will devote more of their resources toward technologies that have a clear, near-term market potential (Goals 1 and 2). They will also have significant efforts directed toward the development of concept and production prototype vehicles. In this regard, PNGV differs from the USABC, where no distinction is made between funding from the government and funding from the automakers. However, DOE funds long-term, high-risk research on batteries under its Exploratory Technology Research program (see Table 2-1).

Both PNGV and the USABC involve the development of precompetitive technologies, but in both cases the three automakers all have related proprietary programs. Thus, Chrysler, Ford, and GM are working independently on EVs, in addition to collaborating on EV battery development through the USABC. Each company participates in the development of advanced automotive technology under PNGV, as well as developing its own proprietary concept vehicle that incorporates its own choice of advanced technologies (by 2000) and a production prototype (by 2004).

The NRC Standing Committee to Review the Research Program of PNGV noted in its second report that some of the technologies under consideration for PNGV would, if implemented, result in very significant changes in the current highway-vehicle-fuel infrastructure (NRC, 1996). The NRC PNGV committee identified the infrastructure issue as an ongoing, integral part of the PNGV program. Major infrastructure changes would also be necessary if EVs were widely adopted. For example, recharging a vehicle at home or at a recharging station would replace trips to the gas station, which would have serious implications for the electric utility and oil industries. The participation of EPRI in the consortium ensures that the impact of EVs on the electric utility industry will be considered, but infrastructure requirements, such as standardized recharging systems and battery recycling, must also be addressed. As noted in Chapter 2, a number of other groups are addressing many of the infrastructure issues. In its RFPI, the USABC also asks that battery developers include environmental and safety issues in manufacturing, recycling, and the use and disposal of proposed batteries.

The NRC PNGV committee found that the PNGV technical road map effectively describes the major technologies, the target performance levels, and the schedule for technology development (NRC, 1996). The present committee believes that a technical road map for battery development and manufacturing, as well as for infrastructure issues, would benefit any follow-on program to the USABC.

A common feature of PNGV and the USABC is that both programs have set extremely challenging goals for performance and cost and very aggressive schedules for meeting them. The NRC PNGV committee noted in its third report that ambitious goals stimulate the rapid development of required technology and that significant improvements in fuel economy may result even if the target of "three times" today's fuel economy is not met (NRC, 1997a). In the case of the development of EV batteries, however, the present committee is not convinced that the USABC program will necessarily have a beneficial effect on the development of a competitive EV, especially if the lithium-polymer long-term battery technology fails to meet expectations. As the GAO noted in its assessment of the program, "the benefits of long-term batteries are clear, but their feasibility is uncertain" (GAO, 1995). Although DOE's participation in the USABC is a valuable addition to the consortium, the committee is concerned that DOE may not maintain a balanced portfolio of short-term development programs (under the USABC) and long-term research on EV batteries. For example, DOE may not be devoting sufficient resources to exploratory research on battery systems that might have longer-term applications in EVs beyond the dates specified by the CARB for market sales of EVs and that might provide a backup technology in case the batteries under development by the USABC do not meet their goals.

Unlike the USABC, the PNGV declaration of intent includes a requirement for unbiased peer review (NRC, 1994):

> An unbiased organization acceptable to both the industry and the government, such as the National Academy of Sciences, will be asked to set up a peer review process to comment on the technologies selected for research and on progress made.

The present committee considers that an objective, external peer review would also be beneficial for the USABC, as well as for any follow-on program involving government-industry collaboration. If a requirement for external peer review had been established at the time DOE agreed to participate in the USABC, it would not only have enhanced the USABC management process, but would also have helped communicate nonproprietary results to a broader technical audience, as has been the case with PNGV. The role of external peer reviews is widely regarded in the engineering and science community as an important component of the effective management of technical programs. In a recent study of energy R&D, for example, the President's Committee of Advisors on Science and Technology (PCAST) called for external peer reviews of DOE programs (PCAST, 1997).

Sematech

Sematech is a nonprofit R&D consortium of U.S. semiconductor manufacturers. Sematech has several features in common with the USABC (Daverse, 1997). Both consortia were stimulated by enabling legislation, and both initially received equal funding from government and industry. Both are focused on midterm development over a three- to five-year timeline, and both have focused on applied, rather than basic, research. In addition, R&D is conducted by supplier companies, rather than by member companies of the consortium.

There is one major difference between Sematech and the USABC, however. Because the Sematech consortium members were threatened by overseas competition, their top management enthusiastically supported the consortium. A sense of urgency based on an imminent threat to existing business was not as strong for the USABC, which may partly explain the low financial stake of the three U.S. automakers in the USABC program. However, some innovative approaches to fuel and emissions announced recently by several Japanese car manufacturers may change this situation. Ford recently announced that it will invest about $420 million in a partnership with Daimler Benz and Ballard Power Systems to develop and manufacture fuel-cell powered cars by 2005 (Detroit News, 1997).

Sematech has been very successful. It has fostered an unprecedented degree of precompetitive cooperation among companies that compete directly with each other in the marketplace. A concrete measure of Sematech's success is that, after 10 years of government co-funding, Sematech is now thriving with only member support. The U.S. semiconductor industry has also increased its worldwide market share, which was one of Sematech's objectives. The GAO found that Sematech not only spent government funds wisely, but also concluded that Sematech should be considered a model for future government-industry partnerships (GAO, 1992).

The Sematech experience shows that an industry-led consortium can be very successful, and its many parallels with the USABC are encouraging for the USABC. Some other aspects of Sematech's experience could also improve USABC's strategies, especially the use of technical road maps. Sematech periodically issues detailed documents describing the future needs of the semiconductor manufacturing industries in all of the limiting technology areas. These documents are available to all interested parties, including suppliers to the industry and potential innovators. In the committee's opinion, a USABC technical road map would have been useful for informing the broader technical community of the consortium's activities and needs (subject to proprietary limitations), as well as a valuable tool for guiding battery development.

FINDINGS

Finding 1. The industry-style management model adopted by the USABC has been effective in focusing development on well defined technical targets. The involvement of managers from Chrysler, Ford, and GM has provided valuable real-world experience, knowledge, and judgment, and under their leadership proven industry practices for project management and cost control have been used effectively.

Finding 2. Even though the USABC program appears to have been managed responsibly, the lack of a true peer review process by independent, outside experts in a program that receives significant federal funding raises concerns about the program's accountability for taxpayer dollars.

Finding 3. The resolution of differences between government and industry practices with respect to contractual and organizational issues drew management attention away from the technical development program in the early stages of the USABC, although there is no evidence that the choices of midterm and long-term battery technologies were adversely affected by the differences in business practices.

Finding 4. DOE has addressed the significant issues raised by the "lessons learned" report from GAO and has acted upon items within its jurisdiction. New procedures implemented by DOE have resulted in more rapid implementation of some government-industry cooperative agreements.

Finding 5. The experience of both Sematech and PNGV has shown the value of a technical road map for guiding cooperative government-industry R&D and communicating technical requirements to potential program participants.

6

Conclusions and Recommendations

This chapter provides the committee's conclusions and recommendations regarding the processes used by the USABC to select and manage R&D projects for EV battery technology and the outcomes of these projects based on information received by the committee as of January 1998. Following an overall perspective on the USABC program, the committee addresses four specific areas: program goals, program management, procurement procedures, and battery technologies. A number of these recommendations are relevant to cooperative government-industry R&D programs in general, including the Advanced Battery Initiative, which has been proposed as a follow-on to the USABC.

OVERALL PERSPECTIVE

Conclusion. Participants in the USABC have made a reasonable effort to develop technology for an extremely demanding application. Despite an overly ambitious schedule, which was imposed by regulatory requirements, the USABC has made significant progress. However, no technology meets all of the midterm goals, although work in progress has a significant chance of meeting both the midterm and commercialization goals early in the next decade, or even before the year 2000. The consortium has effectively brought together battery technologists and experts in EVs in creative, focused partnerships but has missed some opportunities to provide broad technical leadership because timely information (such as summaries of progress or a technical road map) was not made public.

The USABC program has been strongly focused on responding to specific challenges imposed by regulatory agencies. The committee generally agrees with the strategic judgments that have guided the program, the most fundamental of which are (1) that development based on existing science should take precedence over research to create new science and (2) that multiple technologies should be pursued simultaneously. The committee considers that, overall, the participants in the USABC have made reasonable efforts to develop technologies for a very demanding application on a very aggressive schedule. The program has demonstrated that the technical difficulties of developing batteries for EVs constitute a major challenge, perhaps even an insurmountable challenge, if success is measured against the stated long-term performance goals. Even when adequate funding is available for R&D, the USABC goals may not be met. For instance, the USABC would like to have an alternative to the lithium-polymer battery as a long-term option. However, a recent solicitation did not elicit any proposals deemed worthy of funding, even though USABC funds were available.

The USABC may have had no option but to try to meet the implementation date of 1998 mandated by CARB in 1990. However, in the committee's opinion, 1998 was not a realistic date. The first two years of the USABC's existence were consumed with the logistics of creating the organization and its operating principles, defining goals, soliciting proposals, formulating principles governing intellectual property, and negotiating contracts. The committee believes that this necessary work was pursued as efficiently as could be expected and that many lessons were learned that can be used for establishing future government-industry collaborative R&D programs. Nevertheless, by the time the logistics had been established, only three years remained before EVs were scheduled to be ready for the marketplace in 1998. Given the time required for vehicle development, little time was left for battery development. Thus, the USABC's options were restricted. Even with a larger financial commitment, there is no reason to believe that more progress would have been made in the development of the chosen battery technologies.

Nevertheless, the committee respects the role of regulatory initiatives in trying to force the pace of progress in this field because of the significant public benefits that could result from reduced vehicle emissions. In the committee's opinion, an implementation horizon after 2000 would have allowed enough time for a more balanced and effective R&D portfolio on EV batteries, including more opportunities to pursue new science that could improve the performance or cost effectiveness of critical components. Some high-risk investments in exploratory research might have paid off in the long term.

The USABC program has focused largely on development and, except for materials, does not have a research component for the pursuit of new science. The progress on batteries being carried forward into Phase II was partly the result of scientific research on new electrode materials. However, this research was conducted in parallel with the development of large cells, and incorporating the new materials was substantially delayed because of the necessary redesign and retesting of the large cells.

The USABC consortium has brought together automotive and battery manufacturers, electric utilities, and DOE to provide a venue for a focused approach to meeting EV requirements. For the first time, a forum was established for integrating the battery into a vehicle, as well as for evaluating the transportation infrastructure requirements for EVs. The interactions between automobile manufacturers and battery manufacturers have added a valuable dimension of systems integration and testing to battery development programs. A number of groups outside of the USABC are addressing infrastructure issues related to EVs, such as standardized charging systems, the availability of materials for the construction of advanced batteries, and battery disposal and recycling. The committee understands that the USABC considers these issues to be beyond its scope, but they will be critical to meeting the USABC's goals. Therefore, the committee believes the USABC would be justified in paying more attention to infrastructure issues in the remainder of its program.

The committee was surprised to find how little public information has been made available on an ongoing basis from the USABC. In the committee's opinion, proprietary concerns do not necessarily preclude the publication of regular public reports summarizing major progress and concerns. The committee was provided with a draft of a detailed final report on Phase I battery R&D; it is not clear to the committee whether this report will be released to the public (USABC, 1997).[1] In the committee's judgment, this draft report is a good (but belated) report that will not invite the participation of entrepreneurs and small businesses that might have been elicited by regular, timely reports. The lack of regular reporting may well have excluded potential participants from the USABC program.

The lack of public information from the USABC program has also encouraged the perception that "EVs are just around the corner." More public information could have helped to inform the public of the difficulties of developing a commercially viable EV. A USABC technology road map similar to the Semiconductor Industry Association's road map would be a useful guide and would present an accurate picture of the status of EV battery development. A road map would also inform a broader community of potential technology developers and encourage and enable "off road map" innovation.

In the committee's opinion, the USABC had an opportunity and a responsibility to support a climate in which battery technology could move forward on a broad front, not just in large-scale projects directly funded for application in a relatively short time. The USABC could have encouraged creative thinking for the longer term by clearly describing the challenges. The consortium missed this opportunity to provide leadership because it has operated in remarkable isolation. The committee recognizes that sensitive intellectual property in individual projects had to be protected, but the participants in the USABC could have been

[1]The final report on Phase I was submitted by the USABC to DOE in December 1997, a year after the work being reported was completed.

provided with more incentives and encouraged to participate in general technical forums. The USABC partners have a common long-term interest in fostering a healthy industry with multiple suppliers for a given EV battery technology. The USABC could have effectively promoted this interest at very low cost by developing and updating a technology road map and by making concerted efforts to summarize progress regularly on a nonproprietary basis.

By adopting a collegial management model involving program officers from DOE, the USABC was able to take advantage of DOE's extensive expertise with battery systems. However, the committee is concerned that other DOE battery R&D (under the Exploratory Technology Research program or elsewhere in DOE) may have been unwisely reduced to cover federal participation in the USABC. The USABC has a focused development program aimed at one very demanding battery application that might be unrealizable in terms of the defined performance goals. The committee believes that the nation has many different interests in battery science and technology (e.g., for load leveling, solar and wind energy storage, smart buildings with energy-saving features, remote power applications, and portable power sources) and that these interests should be managed, financially and otherwise, in a balanced way by DOE and other federal research programs.

Overall, the 3M/Hydro-Québec lithium-polymer battery program appears to be a good model for future joint development projects between government and industry. In the cooperative environment provided by the USABC, several major companies, including 3M, Hydro-Québec, and the three automotive companies, brought together the needed expertise, which significantly advanced the technology. Participation of national laboratories also contributed to the success of the project. Significant progress has been made toward developing a battery that is expected to meet, or even exceed, the interim commercialization criteria for an EV. The establishment of a U.S.-based manufacturing capability is also well under way. The success of the 3M/Hydro-Québec project is due in large part to the selection of a technology that was mature enough to be developed in the required time frame but not sufficiently mature at the start of the USABC program for collaborative development to be constrained by concerns over sharing proprietary information. Careful project design and management ensured that material selection, cell design, and process development are being done in good order.

Recommendation. In future programs and in current programs outside the USABC, DOE should focus more on research that advances the science of electrochemical systems that have the potential to meet the long-term performance and cost criteria for electric vehicles, as opposed to development of systems based on existing science. A significant portion of the scientific research should be focused on battery technologies that have not been included in the USABC program.

Recommendation. In concert with other groups addressing infrastructure issues, the USABC should use some of its remaining Phase II funds to address infrastructure for its most promising long-term technology, in particular, ensuring the availability of materials and recycling of batteries.

Recommendation. The USABC should communicate its targets and results on a regular basis to the technical community, within the bounds of agreements on proprietary information. The USABC should develop a technical road map to guide battery development, even outside the USABC, and to identify barriers to the implementation of batteries in commercially viable electric vehicles. A technical road map would also be valuable for follow-on programs to the USABC.

Recommendation. Given the trend toward cooperative government-industry technology development and commercialization, DOE should collaborate with other federal agencies to develop a governmentwide approach to the efficient implementation of cooperative agreements. These procedures could also be used for launching a follow-on program to the USABC.

PROGRAM GOALS

Conclusion. The USABC acted appropriately in establishing performance goals consistent with both the CARB mandate and the marketing assumptions made by the three major U.S. automakers. The goals reflect a legitimate view of requirements to meet the CARB mandate with vehicles that could be successful in the marketplace. A successor to the USABC may have the latitude to rethink the performance goals in the light of ongoing technical developments and changing market potential.

A fundamental assumption of the USABC program was that meeting the CARB mandate required that except for vehicle range, EV performance be comparable to the performance of current vehicles powered by gasoline-fueled ICEs. The committee concluded that the approach taken by the USABC was reasonable in light of the regulatory requirements. The USABC relied heavily on its own marketing study and marketing studies by the U.S. automakers to compare the competitiveness of EVs and ICE vehicles. Other marketing strategies, based partly on the advantages of EVs, such as quietness, cleanliness, automatic in-home "refueling," and potentially reduced maintenance, as well as the possibility of an EV being used to complement a household ICE vehicle (e.g., a smaller "neighborhood" vehicle for shorter trips) have been discussed in the literature (see, for example, Kurani et al., 1996). If the USABC had viewed the EV not only as a competitor with the gasoline-fueled ICE vehicle, it might have established more attainable performance goals.

Battery systems developed to meet the midterm goals were not expected to gain a sustainable share of the market for EV batteries or even to benefit the development of battery systems to meet the long-term goals. The committee observed that the main purpose of developing midterm batteries was to meet (or show a good faith effort at meeting) the original CARB mandate (which has since been changed) for the ZEV market share in the late 1990s. The committee questions the wisdom of using short-term regulatory requirements to establish an initial technical direction for the USABC's program.

Cost goals should be a high priority for the USABC program. The projected full cost of ownership and operation of the battery is one very important consideration for EV development. Although the USABC recognizes the importance of the cost issue, it has not used consistent methods or realistic assumptions to address it. For example, a distinction should to be made between battery cost and battery selling price, and models used to predict cost reductions associated with increases in production levels should be explained better. In addition, detailed cost estimates are extremely important for guiding development and setting targets.

Recommendation. Participants in a follow-on program to the USABC should allocate some program funds to examining a broader spectrum of electric vehicle concepts and related market opportunities, in addition to developing a replacement vehicle for a gasoline-fueled ICE vehicle that could capture a significant share of the conventional automobile market.

Recommendation. The participants in a follow-on program to the USABC should make a critical assessment of the economics of batteries for electric vehicles. The issue of cost should be addressed early in the program, and cost projections should be monitored and tested by consistent methods throughout battery development. Meeting cost goals should be an important criterion for maintaining funding.

PROGRAM MANAGEMENT

Conclusion. The three major U.S. automakers have managed the USABC program effectively using proven industry practices and have made appropriate decisions to narrow the portfolio of battery technologies as the program has proceeded.

In the committee's judgment, the industry-led consortium model adopted by the USABC has effectively focused development on well defined, but extremely challenging, technical targets. The three U.S. automakers have managed the program effectively and have made appropriate decisions to narrow the portfolio of battery technologies as the program has proceeded. As the consortium has "matured," the participants have focused on areas where their expertise is most relevant and useful. For example, the national laboratories have now focused primarily on testing and materials development.

The committee observed that both funding and decision making within the consortium are highly leveraged. Although funding levels of the three automakers and EPRI are proprietary, the committee estimates that the contribution of the automotive companies is significantly less than the contribution of DOE, although most of the consortium's management decisions are made by the automakers. The relatively modest financial commitment of the automakers raises some questions about the commitment of high-level managers at Chrysler, Ford, and GM. Nevertheless, the committee recognizes that the automakers have made substantial financial commitments to the development of EVs, especially in the area of vehicle engineering. On the whole, USABC management has been effective in focusing on promising technologies. Also, because the three automotive companies will be customers for EVs and must ensure that batteries developed under the USABC will be compatible with their vehicle requirements, they should have the overall responsibility for their development.

The management processes adopted by the USABC include most proven industry practices. Effective mechanisms have been established to ensure that the statements of work for each project are being followed or are modified in accordance with established procedures. Appropriate mechanisms have also been established for focusing R&D on critical issues, and the implementation of Go/No Go decisions has been informed and decisive. The USABC's management protocol requires that projects meet milestones in order to be continued; renewal is not automatic.

Despite the many excellent features of USABC program management, the committee believes that some changes should be made. The USABC oversight process includes a "peer review" step, but not in the sense that peer review is understood in the context of projects supported by government agencies. USABC peer reviews are done by personnel from within the management structure, rather than independent, outside experts. In the committee's judgment—based on its collective experience of R&D program management—the USABC peer review process is unsatisfactory because it is not sufficiently objective. For example, program managers are likely to defend developers for whom they have oversight responsibility.

A recent report from the NRC (1997b) defines "peer review" as the "use of independent external experts" and argues in favor of using peer reviews for government-supported technology development programs similar to the USABC. The committee believes that peer reviews of the USABC program at regular, specified intervals, and when major new support areas are established, would be useful. Peer reviews would minimize concerns about future decisions regarding technology selection and the financial contributions required of technology developers, as well as the quality of program management. Peer reviews could also provide a different perspective on the technology programs, which would increase the likelihood of finding the optimal midterm and long-term approaches. The committee believes that peer reviews would improve the management processes of the USABC and any follow-on battery initiative.

DOE has responsibilities as a partner and as an overseer in the USABC program. DOE's oversight function is exercised largely through a number of functions laid out in the cooperative agreement between DOE and the USABC. On an operational basis, DOE's participation appears to the committee to be largely collegial rather than directive. In general, this arrangement seems to have worked smoothly and yielded good results. However, it is not a sound practice to have the same individuals fulfill collegial and oversight roles. The DOE must be responsible for ensuring that the public funds spent on the USABC are managed effectively and responsibly.

Recommendation. Within the limits imposed by proprietary considerations, regular peer reviews of the USABC's ongoing programs should be implemented immediately to provide objective assessments and to support decision making. Peer reviews should also be incorporated into any follow-on program that receives federal funding.

Recommendation. DOE should keep its collegial and oversight roles in the USABC program separate by ensuring that managers above the level of participants receive appropriate progress reports.

PROCUREMENT

Conclusion. The USABC has solicited proposals, chosen contractors, and made awards using accepted procurement processes. Although methods for verifying practical accountability have been established, the progress made by the USABC has not been satisfactorily disseminated to a broad audience. As a result, some opportunities for technical advancement have been lost. The USABC's policies on intellectual property and its processes for evaluating proposals may have precluded the participation of small companies and international partners, especially organizations with relatively mature battery systems.

In the committee's judgment, the USABC has used a sound, effective process to solicit proposals. The original RFPI was broad and open, and the USABC stimulated an impressive amount of creative collaborative activity among technology developers. For example, the joint project between 3M and Hydro-Québec, which was brokered by the USABC, appears to be a successful collaboration between two owners of potentially valuable, complementary technologies. The result of this collaboration appears to be much greater than the sum of its parts.

Many small companies, however, were unable to compete for contracts effectively because the USABC's proposal evaluation criteria placed considerable emphasis on the developer having an established large-scale manufacturing capability. For example, the circumstances of small companies working to develop battery technology relevant to EVs were apparently not taken into account.[2] In addition, the fundamental objective of fostering American competitiveness in the automobile and battery industries tends to limit international collaboration. In recognition of the need for practical technologies, Phase II was designed to counter this tendency by soliciting proposals from Asian organizations. No new international partnerships have been developed, however.

The USABC's policies governing intellectual property favor the consortium partners rather than the developers of the technology. Other organizations, such as Sematech, which have adopted policies that are more favorable to the developers, seem more likely to promote the enthusiastic development of desired technologies. In this respect, the operational practices of the USABC appear to be defensive. The RFPI requirement that companies reveal commercially sensitive information almost certainly has discouraged battery developers with relatively mature systems—which they had developed over many years at considerable expense—from participating.

The committee applauds the consortium's requirement that testable items, particularly cells, batteries, and packs, be delivered periodically. The committee also believes that the national laboratories have performed a valuable service as independent evaluators of technology, especially through testing programs tailored to the needs of the USABC. By this mechanism, the consortium has created an excellent method of establishing practical accountability. However, the virtual absence of timely progress summaries for a broader audience is a major deficiency. Proper documentation is not only a means of tracking progress, but is also a valuable tool for analyzing programs and stimulating researchers and vendors. The USABC has probably lost opportunities for technical advancement because of the lack of timely documentation.

BATTERY TECHNOLOGIES

Conclusion. Given the constraints of the CARB mandate, the USABC made a reasonable selection of prospective technologies for EVs and pursued these technologies with appropriate partners. Substantial progress has been made even though none of the required technologies and none of the battery systems currently funded meets the USABC's cost or life objectives.

At the outset of the USABC program, the technologies to support the program's goals did not exist. Although the necessary technologies have still not been developed to the point of meeting these goals, substantial progress has been made. In the committee's judgment, given the established performance goals and timetable of the CARB mandate, the USABC made a reasonable selection of technologies and pursued them with appropriate partners.

[2]As discussed in Chapter 3, however, DOE has created opportunities for small companies through the Small Business Innovative Research program.

The USABC has made substantial progress in the development of Ni/MH (nickel metal hydride) and lithium battery technologies for EV applications and has improved the understanding of design, manufacturing processes, system integration, and cost issues. The committee commends the USABC on developing two competing Ni/MH technologies based on AB_2- and AB_5-type hydrogen absorbing alloys. The USABC now has a fall-back position in the event that one of the battery systems fails to meet program objectives. Unfortunately, recent USABC solicitations have not led to a potential alternative technology for the lithium-polymer battery as a long-term option.

In the committee's judgment, the USABC acted appropriately in closing down unpromising lines of investigation. The consortium made the correct decisions in canceling the sodium-sulfur, lithium-iron disulfide, and lithium-ion polymer battery projects that did not meet critical performance criteria or offer promising business opportunities at the end of Phase I. The committee also considers that the USABC made the correct decision in focusing Phase II on reducing battery costs. None of the battery systems presently funded has met its cost objectives, which will be critical to the successful commercialization of EVs.

The SAFT/ANL lithium-iron disulfide battery program could have benefited from focusing more attention on the well known problems of cell cycle life and calendar life. Calendar life is often the most important parameter because corrosion processes occur at operating temperature regardless of whether cycling is being performed or not. Changes in size that occur with cycling in solid electrode forms can also affect calendar life. With more emphasis on corrosion and seal behavior early in the project, a more cost-effective assessment of the technical barriers to developing a lithium-iron disulfide battery for EV applications could have been made. The SAFT program was eventually terminated.

After lead-acid battery technology was excluded from the USABC portfolio, the manufacturers of lead-acid batteries and their suppliers decided to form their own consortium and undertake an applied research program to make their technology competitive with battery technologies developed under the USABC. The ALABC (Advanced Lead Acid Battery Consortium) program has made significant technical progress with very little government funding.

In spite of the progress made by the USABC, much remains to be done in the areas of battery systems integration and testing, reliability, and infrastructure development before commercialization can become a reality. Safety remains a concern for all battery systems, but particularly for the Duracell/VARTA lithium-ion battery with an organic electrolyte. Nissan's plans to market EVs incorporating Sony lithium-ion technology have apparently been delayed by major safety issues (Wilkinson, 1997). Cycle and calendar life have yet to be determined for most of the battery systems. The Duracell/VARTA program appears to be a good model for the systems testing and integration of modules that will be required for the 3M/Hydro-Québec technology. Experienced battery manufacturers, working with the automotive companies, have produced a project model that can be adapted to other projects.

Three areas may require additional funding in the future in the 3M/Hydro-Québec project: the recycling and disposal of spent batteries, safety testing of used cells nearing the end of life, and the evaluation of performance under extreme temperatures and stressful-use conditions.

The committee questions the continued development of Ni/MH technology for application in ZEVs because current estimates suggest that Ni/MH batteries will not approach performance goals. The Ni/MH battery, which does not meet the midterm performance or cost requirements, will be discontinued after cost reduction efforts are completed in 1998 (Heitner, 1998). It will remain under development in a separate program to meet high-power energy storage requirements for hybrid vehicles for the PNGV program.

The USABC established midterm technology programs in response to the CARB requirement of 2 percent ZEVs in 1998. However, CARB has since changed the mandate to 10 percent ZEVs in 2003 and is considering revising the requirement to include hybrid vehicles as ZEVs. The committee, therefore, believes that the USABC is justified in reevaluating its midterm battery programs. The Duracell/VARTA lithium-ion battery meets the midterm performance criteria but not the cost requirements. The necessity of redesigning the battery and/or using modified materials suggests that the program may not meet the schedule required for midterm batteries. Because the final design freeze for the Duracell/VARTA battery is scheduled at the same time as the design freeze for the 3M/Hydro-Québec battery, the USABC should consider whether the Duracell/VARTA battery could be developed to meet the interim commercialization goals and, if so, what additional work would be required.

Recommendation. In Phase II, the USABC should place more emphasis on evaluating promising battery systems under extreme temperatures and stressful-use conditions. Batteries should also be subjected to extensive safety and reliability testing from the beginning to the end of life.

Recommendation. The USABC should reevaluate its midterm battery programs. The development of the lithium-ion battery should be continued if it has the potential to meet the interim commercialization criteria.

References

Abacus Technology Corporation. 1993. Lessons Learned under the United States Advanced Battery Consortium, Final Report. November 8. Prepared for the U.S. Department of Energy, Office of Transportation Technologies. Chevy Chase, Md.: Abacus Technology Corporation.

Abacus Technology Corporation. 1996. Review of the U.S. Department of Energy's Implementation of the 1993 USABC Lessons Learned Study. May 17. Prepared for the U.S. Department of Energy, Office of Transportation Technologies. Chevy Chase, Md.: Abacus Technology Corporation.

Branscomb, L.M., and J.H. Keller. 1998. Investing in Innovation: Creating a Research and Innovation Policy That Works. Cambridge, Mass.: Massachusetts Institute of Technology Press.

CARB (California Air Resources Board). 1967. The California Air Resources Board. Sacramento, Calif.: California Air Resources Board.

CARB. 1994. 1994 Low-Emission Vehicle and Zero-Emission Vehicle Program Review. Sacramento, Calif.: California Air Resources Board, Mobile Source Division.

CARB. 1996. The California Zero-Emission Vehicle Program. Sacramento, Calif.: California Air Resources Board.

Christensen, C.M. 1997. The Innovator's Dilemma: When New Technologies Cause Great Firms to Fail. Boston, Mass.: Harvard Business School Press.

Daverse, C. 1997. Sematech: an alternative model for a government/industry R&D consortium. Presentation to the Committee to Review the U.S. Advanced Battery Consortium's Electric Vehicle Battery R&D Selection Process, National Research Council, Washington, D.C., December 10, 1997.

Davis, R. 1997. Establishment of technical goals and objectives for battery development. Presentation to the Committee to Review the U.S. Advanced Battery Consortium's Electric Vehicle Battery R&D Project Selection Process, National Research Council, Washington, D.C., October 20, 1997.

Davis, S.C. 1997. Transportation Energy Databook (17th ed.). ORNL-6919. Prepared for the U.S. Department of Energy, Office of Transportation Technologies. Oak Ridge, Tenn.: Oak Ridge National Laboratory.

Detroit News. 1997. Ford, Daimler unite to develop fuel cells. December 16, 1997.

DOE (U.S. Department of Energy). 1991. DOE-USABC Cooperative Agreement, Advanced Battery Research and Development for Electric Vehicles. Washington, D.C.: U.S. Department of Energy, Office of Advanced Automotive Technology.

DOE. 1996. New USABC Battery Commercialization Criteria. Presented to the U.S. Department of Energy's Automotive Technology Development Contractors' Coordination Meeting, Detroit, Michigan, October 30, 1996.

GAO (General Accounting Office). 1992. Lessons Learned from Sematech. Washington, D.C.: U.S. General Accounting Office.

GAO. 1994. Technology Transfers: Benefits of Cooperative R&D Agreements. Washington, D.C.: U.S. General Accounting Office.

GAO. 1995. Electric Vehicles. Efforts to Complete Advanced Battery Development Will Require More Time and Funding. Report to the Ranking Minority Member, Committee on Governmental Affairs, GAO/RCED-95-234. Washington, D.C.: U.S. General Accounting Office.

Heitner, K. 1997. DOE role in USABC management and battery technology selection. Presentation to the Committee to Review the U.S. Advanced Battery Consortium's Electric Vehicle Battery R&D Project Selection Process. National Research Council, Washington, D.C., December 10, 1997.

Heitner, K. 1998. Personal communication. Office of Advanced Automotive Technology, U.S. Department of Energy, Washington, D.C.

Kalhammer, F., A. Kozawa, C.B. Moyer, and B.B. Owen. 1995. Performance and Availability of Batteries for Electric Vehicles: A Report of the Battery Technical Advisory Panel (December 11). Prepared for the California Air Resources Board, El Monte, California.

Kurani, K., D. Sperling, and T. Turrentine. 1996. The marketability of electric vehicles: battery performance and consumer demand for driving range. Pp. 153–158 in Proceedings of the 11th Annual Battery Conference on Applications and Advances, Long Beach, California. January 9(12, 1996. Piscataway, N.J.: Institute for Electrical and Electronic Engineers.

Leclair, V. 1997. EPA proposes a national low-emission vehicle program. Environmental Science and Technology 31(7): 310A–311A.

Letourneau, C., M. Gauthier, A. Belanger, D. Kuller, and J. Hoffman. 1997. Lithium polymer pack design. Presented at EVS-14, the 14th International Electric Vehicle Symposium, Orlando, Florida, December 15–17, 1997.

Moseley, P. 1997. Advanced Lead-Acid Battery Consortium. Presentation to the Committee to Review the U.S. Advanced Battery Consortium's Electric Vehicle Battery R&D Project Selection Process, National Research Council, Washington, D.C., October 20, 1997.

Nichols, R. 1992. Advanced Battery Consortium. Testimony before the Committee on Government Affairs. United States Senate, 102nd Congress, Second Session, February 18, 1992. Washington, D.C.: U.S. Government Printing Office.

NRC (National Research Council). 1994. Review of the Research Program of the Partnership for a New Generation of Vehicles. Board on Energy and Environmental Systems and Transportation Research Board. Washington, D.C.: National Academy Press.

NRC. 1996. Review of the Research Program of the Partnership for a New Generation of Vehicles, Second Report. Board on Energy and Environmental Systems and Transportation Research Board. Washington, D.C.: National Academy Press.

NRC. 1997a. Review of the Research Program of the Partnership for a New Generation of Vehicles, Third Report. Board on Energy and Environmental Systems and Transportation Research Board. Washington, D.C.: National Academy Press.

NRC. 1997b. Peer Review in the Department of Energy-Office of Science and Technology: Interim Report. Washington, D.C.: National Academy Press.

Patil, P. 1997. Sponsor expectations for study. Presentation to the Committee to Review the U.S. Advanced Battery Consortium's Electric Vehicle Battery R&D Project Selection Process, National Research Council, Washington, D.C., October 20, 1997.

PCAST (President's Committee of Advisors on Science and Technology). 1997. Federal Energy Research and Development of the Challenges of the Twenty-First Century (November 5). Washington, D.C.: Executive Office of the President.

Rauhe, B. 1997. In-depth review of a nickel-metal hydride battery contract. Presentation to the Committee to Review the U.S. Advanced Battery Consortium's Electric Vehicle Battery R&D Project Selection Process, National Research Council, Washington, D.C., October 20, 1997.

Stroven, M. 1997. Processes used by USABC to solicit proposals, choose contractors, and make awards. Presentation to the Committee to Review the U.S. Advanced Battery Consortium's Electric Vehicle Battery R&D Project Selection Process, National Research Council, Washington, D.C., October 19, 1997.

Sutula, R. 1997. Letter to Dr. Jill Wilson, National Research Council staff. U.S. Department of Energy Office of Advanced Automotive Technology. Washington, D.C., November 6, 1997.

Thorpe, N. 1997. Introduction to USABC. Presentation to the Committee to Review the U.S. Advanced Battery Consortium's Electric Vehicle Battery R&D Project Selection Process, National Research Council, Washington, D.C., October 20, 1997.

USABC (United States Advanced Battery Consortium). 1991. Advanced Battery Technology Request for Proposal Information (RFPI), April 4, 1991. Dearborn, Mich.: United States Advanced Battery Consortium.

USABC. 1996. USABC Moves into Phase II, EPRI Extends Commitment. USABC: News for and about the United States Advanced Battery Consortium. 1996 (summer):1.

USABC. 1997. Final Report: Phase I Battery Research and Development 1992–1996. Submitted to the U.S. Department of Energy Advanced Battery Consortium. December. Dearborn, Mich.: USABC.

Wells, J. 1993. Technology Transfer: Implementation of CRADAs at NIST, Army, and DOE. Testimony before the Subcommittee on Energy, Committee on Science, Space, and Technology, House of Representatives, GAO/T-RCED-93-53, June 10, 1993.

Wilkinson, S.L. 1997. Electric vehicles gear up. Chemical and Engineering News 75(41): 18–24.

APPENDICES

APPENDIX A

Biographical Sketches of Committee Members

Larry R. Faulkner (*chair*) is professor of chemistry and president of the University of Texas at Austin. His previous positions include assistant professor at Harvard University, professor of chemistry at the University of Texas at Austin, and professor of chemistry, head of the Department of Chemistry, dean of the College of Liberal Arts and Sciences, and provost and vice chancellor for academic affairs at the University of Illinois at Urbana-Champaign. Dr. Faulkner is a former president and vice president of the Electrochemical Society and chair of the Executive Committee of the Electrochemical Society's Physical Electrochemistry Division. He has also served as vice president of the International Society of Electrochemistry and is a fellow of the American Association for the Advancement of Science. Dr. Faulkner has served on a number of National Research Council committees, including the Committee on Electrochemical Aspects of Energy Conservation and Production, the Air Force Panel on Basic Research, and the Committee on Critical Technologies: The Role of Chemistry and Chemical Engineering in Maintaining and Strengthening American Technology. He received the U.S. Department of Energy Award for Outstanding Scientific Achievement in Materials Chemistry and the American Chemical Society's Award in Analytical Chemistry. His areas of expertise include chemiluminescence, electron transfer mechanisms, assemblage of chemical devices from thin films of molecular materials, molecular organization, photoredox processes in zeolites, and electrochemical instrumentation. He received a B.S. degree from Southern Methodist University and a Ph.D. from the University of Texas at Austin.

Kathryn R. Bullock is development manager, Power Sources, at Medtronic, Inc., where she is responsible for developing lithium batteries and other power sources for implantable medical devices. Before joining Medtronic, she was technical manager, Batteries and Purchased Products, and a member of the Energy Systems Management Team at AT&T Bell Laboratories with responsibility for the development and engineering of batteries, cables, and other electronic products for telecommunications power systems. From 1980 to 1991, Dr. Bullock was manager of the Chemical Research Department of Johnson Controls, Inc., where she supervised the research and development (R&D) of new battery systems, innovative processes and designs for lead-acid batteries, and battery materials. She is a fellow of the Royal Society of Chemistry (London) and of the American Institute of Chemists and has served as president and vice president of the Electrochemical Society. Dr. Bullock received a B.A. degree in English from the University of Colorado and a Ph.D. in physical chemistry from Northwestern University.

Paul A. Kohl is professor and institute fellow of the School of Chemical Engineering, Georgia Institute of Technology. He was formerly a technical supervisor at AT&T Bell Laboratories, where his activities included electrochemical processing, the establishment of a thin film analytical facility, and the development of processes for metal/polymer deposition on silicon multichip modules. Dr. Kohl's current research interests include the electrochemistry of semiconductors, batteries and deposition, and semiconductor processing. He has published more than 60 papers in scientific journals, and his research has resulted in 22 patents in electrochemistry and semiconductor processes. He is editor of the *Journal of the Electrochemical Society* and was founding editor of *The Electrochemical Society Interface*. Dr. Kohl received a B.S. from Bethany College and a Ph.D. from the University of Texas at Austin, both in chemistry.

Carl A. Kukkonen is director of the Center for Space Microelectronics Technology and manager of the Supercomputing Project at the Jet Propulsion Laboratory (JPL), California Institute of Technology. He is responsible for the management and technical leadership of programs to develop and demonstrate advanced technology for National Aeronautics and Space Administration and the U.S. Department of Defense and to deliver this technology to industry for commercialization. Before joining JPL, Dr. Kukkonen spent eight years with Ford Motor Company, where he conducted a technological assessment of the potential for using hydrogen as an alternative automotive fuel. He was also project leader of the team that designed, developed, and demonstrated the first direct injection passenger car diesel engine. Dr. Kukkonen received a B.S. degree in physics from the University of California, Davis, and M.S. and Ph.D. degrees, also in physics, from Cornell University.

Alexander MacLachlan (NAE) retired as deputy undersecretary for R&D management at the U.S. Department of Energy (DOE) in 1996. During his two-year tenure at DOE, Dr. MacLachlan was responsible for oversight of partnership activities between the national laboratories and private industry, and he assisted in reengineering the DOE's research management structure. In 1993, Dr. MacLachlan retired as senior vice president for R&D at DuPont after more than 36 years of service. He has served as a director of the Industrial Research Institute and as a member of the Fermi Board of Overseers and is currently a trustee of the Bartol Institute at the University of Delaware and the Mt. Cuba Astronomical Observatory. Dr. MacLachlan is a member of the Secretary of Energy's External Advisory Board and of the Sandia President's Advisory Council. He was elected to the National Academy of Engineering in 1992. He received a B.S. in chemistry from Tufts University and a Ph.D. in physical organic chemistry from the Massachusetts Institute of Technology.

James A. McIntyre is a research scientist in the Designed Products and Devices Laboratory at Dow Chemical Company's central research and development facility in Midland, Michigan. His areas of expertise include electrochemical techniques for waste stream reduction, fuel cells, electroactive materials, and applications of oxygen electrochemistry to energy reduction. He is a member of the Electrochemical Society and the American Association for the Advancement of Science. Dr. McIntyre received B.Sc. and M.Sc. degrees in chemistry from the University of Manitoba and a Ph.D. in physical chemistry from Rensselaer Polytechnic Institute.

Barry Miller is Frank Hovorka Professor of Chemistry at Case Western Reserve University. His previous positions include member of the technical staff at AT&T Bell Laboratories and instructor of chemistry at Harvard University. Dr. Miller's research interests include interfacial kinetics, electronic materials processing, electrosynthesis (semiconductors, polymers, and superconductors), and the development of new instrumental methods in electrochemistry. He received the Charles N. Reilly Award of the Society for Electroanalytical Chemistry in 1994 and the David C. Grahame Award in Physical Electrochemistry from the Electrochemical Society in 1991. Dr. Miller is president of the Electrochemical Society for 1997–1998 and has served in several other positions in the Electrochemical Society. He is former president of the Society for Electroanalytical Chemistry and former United States national secretary of the International Society for Electrochemistry. Dr. Miller received an A.B. from Princeton University and a Ph.D. degree from the Massachusetts Institute of Technology, both in chemistry.

David L. Morrison is retired director of the Office of Nuclear Regulatory Research, U.S. Nuclear Regulatory Commission. His previous positions include technical director of the Energy, Resource and Environmental Systems Division,

MITRE Corporation; president of the IIT Research Institute; and director of program development and management, Battelle Memorial Institute. He has been a member of the NRC's Energy Engineering and National Materials Advisory Boards, has chaired the NRC Committee on Alternative Energy R&D Strategies, and has served on the NRC Committee on Fuel Economy of Automobiles and Light Trucks. His areas of expertise include research management, energy and environmental research, nuclear chemistry, and physical chemistry. Dr. Morrison has a B.S. degree from Grove City College and a Ph.D. in chemistry from the Carnegie Institute of Technology.

Brijesh Vyas is technical manager of the Energy Conversion Technology Group at Bell Laboratories-Lucent Technologies, Murray Hill, New Jersey, where he is responsible for research on advanced materials and technologies for high energy density batteries. He has led efforts to develop rechargeable lithium batteries and use nickel-cadmium and nickel-hydrogen batteries for satellite applications. Dr. Vyas is responsible for technology transfer to manufacturing and the liaison with battery users for alternative energy solutions. He was formerly on the staff at Brookhaven National Laboratory and has been a guest professor at the Technical University of Denmark in Copenhagen. He received the Sam Tour Award from the American Society of Materials and Testing in 1983. His areas of expertise include materials, electrochemistry, and corrosion. Dr. Vyas received a BTech degree in metallurgical engineering from the Indian Institute of Technology in Bombay and a Ph.D. in materials science from the State University of New York, Stony Brook.

Robert D. Weaver is retired from the Electric Power Research Institute (EPRI), where he was manager for more than 15 years of a variety of battery R&D contracts, especially relating to high-energy fused-salt load-leveling systems. Before joining EPRI, he was responsible for general electrochemical and client-supported research, including work on various battery systems, at Stanford Research Institute (SRI) International. Mr. Weaver was a researcher with General Motors Corporation (Delco Remy and Defense Research Laboratories) from 1958 to 1967, where he performed research leading to new battery systems for electric vehicles and space power systems. He was a civilian research chemist at the Michelson Laboratory of the U.S. Naval Ordnance Test Station at China Lake, California, from 1953 to 1958, where he conducted research on electro-organic reactions. Mr. Weaver has many publications and five patents in batteries and fuel cells. Much of his work has been in the field of high-energy batteries using fused salts. He is currently listed as a scientific fellow at SRI International. Mr. Weaver has a B.A. in chemistry and mathematics from Blackburn University and an M.S. in chemistry from Kansas State College.

APPENDIX
B

Committee Meetings and Other Activities

FIRST COMMITTEE MEETING, OCTOBER 19–21, 1997
WASHINGTON, D.C.

The following presentations were made to the committee:

Sponsor expectations for the study
Pandit Patil, Office of Advanced Automotive Technologies (OAAT), U.S. Department of Energy (DOE)

Introduction to the USABC
Norm Thorpe, USABC/General Motors

Establishment of technical goals and objectives for battery development
Bob Davis, USABC/Chrysler

Processes used by USABC to solicit proposals, choose contractors, and make awards
Marlyn Stroven, USABC/Ford

Measurement and evaluation of contractor performance by USABC
Jim Pass, USABC/General Motors

In-depth review of a nickel-metal hydride battery contract
Bruce Rauhe, USABC/Electric Power Research Institute

Summary of major Phase I contracts
Jim Pass, USABC/General Motors
Marlyn Stroven, USABC/Ford

USABC plans and activities for Phase II
Marlyn Stroven, USABC/Ford

General Accounting Office (GAO) assessment of USABC activities
Susan Kladiva, GAO

Advanced Lead-Acid Battery Consortium (ALABC)
Pat Moseley, ALABC

SECOND COMMITTEE MEETING, DECEMBER 10–12, 1997 WASHINGTON, D.C.

The following presentations were made to the committee:

Sematech: An alternative model for a government-industry R&D consortium
Chris Daverse, Sematech

Lithium battery technologies
Brijesh Vyas, committee member

Development of lithium-polymer batteries under contract to USABC
Chuck Donnelly, 3M
Claude Létourneau, Argo-Tech Productions, Inc.

DOE role in USABC management and battery technology selection
Ken Heitner, OAAT, DOE

SITE VISIT TO 3M, JANUARY 14, 1998 MINNEAPOLIS, MINNESOTA

A representative of the committee visited 3M to see lithium-polymer battery hardware developed by 3M and Hydro-Québec under the USABC program.

APPENDIX C

Acronyms

ALABC	Advanced Lead Acid Battery Consortium
ANL	Argonne National Laboratory
CARB	California Air Resources Board
CRADA	cooperative research and development agreement
DOE	U.S. Department of Energy
EPACT	Energy Policy Act of 1992
EPRI	Electric Power Research Institute
EV	electric vehicle
ETR	exploratory technology research
FAR	Federal Acquisition Regulation
FMEA	failure modes and effects analysis
FY	fiscal year
GAO	General Accounting Office
GM	General Motors Corporation
GMO	GM-Ovonic
ICE	internal combustion engine
JPL	Jet Propulsion Laboratory

Li/FeS_2	lithium-iron disulfide
Na/S	sodium-sulfur
Ni/MH	nickel metal hydride
Ni/Cd	nickel cadmium
NRC	National Research Council
OAAT	Office of Advanced Automotive Technologies (U.S. Department of Energy)
OTT	Office of Transportation Technologies (U.S. Department of Energy)
PNGV	Partnership for a New Generation of Vehicles
R&D	research and development
RFPI	request for proposal information
TAC	technical advisory committee
USABC	United States Advanced Battery Consortium
USCAR	United States Council for Automotive Research
ZEV	zero emission vehicle